D0742774

From the

WANDERING JEW

to

WILLIAM F.
BUCKLEY JR.

ALSO BY MARTIN GARDNER

MARTIN GARDNER

on science, literature, and religion

From the

WANDERING JEW

to

WILLIAM F. BUCKLEY JR.

 Prometheus Books
59 John Glenn Drive
Amherst, New York 14228-2197

Published 2000 by Prometheus Books

Inquiries should be addressed to
Prometheus Books
59 John Glenn Drive
Amherst, New York 14228–2197
VOICE: 716–691–0133, ext. 207
FAX: 716–564–2711
WWW.PROMETHEUSBOOKS.COM

04 03 02 01 00   5 4 3 2 1

Library of Congress Cataloging-in-Publication Data

Gardner, Martin, 1914–
From the wandering Jew to William F. Buckley Jr. : on science, literature, and religion / Martin Gardner.
    p. cm.
    ISBN 1–57392–852–6 (alk. paper)
    I. Title.

AC8 .G333 2000
081—dc21   00–055277
                                                        CIP

Printed in the United States of America on acid-free paper

CONTENTS

PART 2. BOOK REVIEWS

INTRODUCTION

From time to time, when the impulse seizes me, I sound off about topics of interest to me, and respond to requests for book reviews if I think I have something significant to say about a book. The columns I wrote about mathematics for *Scientific American* are gathered in fifteen books. Five Prometheus books are collections of columns about science and pseudoscience that I have been contributing for many years to the *Skeptical Inquirer*. Essays and reviews that fall outside those two categories have been put into three books: *Order and Surprise*, *Gardner's Whys & Wherefores*, and *The Night Is Large*. This is the fourth such collection.

As you will see, the topics of this rambling anthology are mainly attacks on bogus science and what I regard as religious superstition. Two reviews are of books about Lewis Carroll, one is about a forgotten children's magazine, two are attacks on antirealist philosophies of mathematics, and one is a collection of introductions to three books by H. G. Wells.

The long essay on Hugo Gernsback, whose magazine *Science and Invention* was one of the great delights of my boyhood, is published here for the first time.

Martin Gardner

ESSAYS

THE WANDERING JEW
AND THE
SECOND COMING

For the son of man shall come in the glory of his Father with his angels; and then he shall reward every man according to his works. Verily I say unto you, There be some standing here, which shall not taste of death, till they see the Son of man coming in his kingdom.

Matthew 16: 27, 28

The statement of Jesus quoted above from Matthew, and repeated in similar words by Mark (8:38, 9:1) and Luke (9:26, 27) is for Bible fundamentalists one of the most troublesome of all New Testament passages.

It is possible, of course, that Jesus never spoke those sentences, but all scholars agree that the first-century Christians expected the Second Coming in their lifetimes. In Matthew 24, after describing dramatic signs of his imminent

---

This essay first appeared in *Free Inquiry* (summer 1995).

return, such as the falling of stars and the darkening of the moon and sun, Jesus added: "Verily I Say unto you. This generation shall not pass until all these things be fulfilled."

Until about 1933 Seventh-day Adventists had a clever way of rationalizing this prophecy. They argued that a spectacular meteor shower of 1833 was the falling of the stars, and that there was a mysterious darkening of sun and moon in the United States in 1870. Jesus meant that a future generation witnessing these celestial events would be the one to experience his Second Coming.

For almost a hundred years Adventist preachers and writers of books assured the world that Jesus would return within the lifetimes of some who had seen the great meteor shower of 1833. After 1933 passed, the church gradually abandoned this interpretation of Jesus' words. Few of today's faithful are even aware that their church once trumpeted such a view. Although Adventists still believe Jesus will return very soon, they no longer set conditions for an approximate date.

How do they explain the statements of Jesus quoted in the epigraph? Following the lead of Saint Augustine and other early Christian commentators, they take the promise to refer to Christ's Transfiguration. Ellen White, the prophetess who with her husband founded Seventh-day Adventism, said it this way in her life of Jesus, *The Desire of Ages*: "The Savior's promise to the disciples was now fulfilled. Upon the mount the future kingdom of glory was represented in miniature. . . ."

Hundreds of Adventist sects since the time of Jesus, starting with the Montanists of the second century, have all interpreted Jesus' prophetic statements about his return to refer to *their* generation. Apocalyptic excitement surged as the year 1000 approached. Similar excitement is now gathering momentum as the year 2000 draws near. Expectation of the Second Coming is not confined to Adventist sects. Fundamentalists in mainstream Protestant denominations are increasingly stressing the imminence of Jesus's return. Baptist Billy Graham, for example, regularly warns of the approaching battle of Armageddon and the appearance of the Antichrist. He likes to emphasize the Bible's assertion that the Second Coming will occur after the gospel is preached to all nations. This could not take place, Graham insists, until the rise of radio and television.

Preacher Jerry Falwell is so convinced that he will soon be raptured—caught up in the air to meet the return of Jesus—that he once said he has no plans for a burial plot. Austin Miles, who once worked for Pat Robertson, reveals in his book *Don't Call Me Brother* (1989) that Pat once seriously considered plans to televise the Lord's appearance in the skies! Today's top native drumbeater for a soon Second Coming is Hal Lindsay. His many books on the topic, starting with *The Late Great Planet Earth*, have sold by the millions.

For the past two thousand years individuals and sects have been setting dates for the Second Coming. When the Lord fails to show, there is often no recognition of total failure. Instead, errors are found in the calculations and new dates set. In New Harmony, Indiana, an Adventist sect called the Rappites was established by George Rapp. When he became ill he said that, were he not absolutely certain the Lord intended him and his flock to witness the return of Jesus, he would think this was his last hour. So saying, he died.

The Catholic church, following Augustine, long ago moved the Second Coming far into the future at some unspecified date. Liberal Protestants have tended to take the Second Coming as little more than a metaphor for the gradual establishment of peace and justice on Earth. Julia Ward Howe, a Unitarian minister, had this interpretation in mind when she began her famous "Battle Hymn of the Republic" with "Mine eyes have seen the glory of the coming of the Lord. . . ." Protestant fundamentalists, on the other hand, believe that Jesus described actual historical events that would precede his literal return to Earth to banish Satan and judge the quick and the dead. They also find it unthinkable that the Lord could have blundered about the time of his Second Coming.

The difficulty in interpreting Jesus' statement about some of his listeners not tasting of death until he returned is that he described the event in exactly the same phrases he used in Matthew 24. He clearly was not there referring to his transfiguration, or perhaps (as another "out" has it) to the fact that his kingdom would soon be established by the formation of the early church. Assuming that Jesus meant exactly what he said, and that he was not mistaken, how can his promise be unambiguously justified?

Emily Dickinson, in one of her poems, said it this way:

> I say to you, said Jesus—
> That there be standing here—
> A Sort, that shall not taste of Death—
> If Jesus was sincere—

During the Middle Ages several wonderful legends arose to preserve the accuracy of Jesus' prophecies. Some were based on John 21. When Jesus said to Peter, "Follow me," Peter noticed John walking behind him and asked, "Lord, what shall this man do?" The Lord's enigmatic answer was, "If I will that he tarry till I come, what is that to thee?"

We are told that this led to a rumor that John would not die. However, the writer of the Fourth Gospel adds: "Yet Jesus said not unto him, he shall not die; but if I will that he tarry till I come, what is that to thee?" Theologians in the Middle Ages speculated that perhaps John did not die. He was either wandering about the earth, or perhaps he ascended bodily into heaven. A more popular legend was that John had been buried in a state of suspended animation, his heart faintly throbbing, to remain in this unknown grave until Jesus returns.

These speculations about John rapidly faded as a new

and more powerful legend slowly took shape. Perhaps Jesus was not referring to John when he said he could ask someone to tarry, but to someone else. This would also explain the remarks quoted in the epigraph. Someone not mentioned in the Gospels, alive in Jesus' day, was somehow cursed to remain alive for centuries until Judgment Day, wandering over the Earth and longing for death.

Who was this Wandering Jew? Some said it was Malchus, whose ear Peter sliced off. Others thought it might be the impenitent thief who was crucified beside Jesus. Maybe it was Pilate, or one of Pilate's servants. The version that became dominant identified the Wandering Jew as a shopkeeper—his name varied—who watched Jesus go by his doorstep, staggering under the weight of the cross he carried. Seeing how slowly and painfully the Lord walked, the man struck Jesus on the back, urging him to go faster. "I go," Jesus replied, "but you will tarry until I return."

As punishment for his rudeness, the shopkeeper's doom is to wander the Earth, longing desperately to die but unable to do so. In some versions of the legend, he stays the same age. In others, he repeatedly reaches old age only to be restored over and over again to his youth. The legend seems to have first been recorded in England in the thirteenth century before it rapidly spread throughout Europe. It received an enormous boost in the early seventeenth century when a pamphlet appeared in Germany about a Jewish shoemaker named Ahasuerus who claimed to be the Wanderer. The pamphlet was endlessly reprinted in Germany and translated into other languages. The result was a mania comparable to today's manias for seeing UFOs, Abominable Snowmen, and Elvis Presley. Scores of persons claiming to be the Wandering Jew turned up in cities all over England and Europe during the next two centuries. In the United States as late as 1868 a Wandering Jew popped up in Salt Lake City, home of the Mormon Adventist sect. It is impossible now to decide in individual cases whether these were rumors, hoaxes by impostors, or cases of self-deceived psychotics.

The Wandering Jew became a favorite topic for hundreds of poems, novels, and plays, especially in Germany where such works continue to proliferate to this day. Even Goethe intended to write an epic about the Wanderer, but he only finished a few fragments. It is not hard to understand how anti-Semites in Germany and elsewhere would see the cobbler as representing all of Israel, its people under God's condemnation for having rejected his Son as their Messiah.

Gustave Doré produced twelve remarkable woodcuts depicting episodes in the Wanderer's life. They were first published in Paris in 1856 to accompany a poem by Pierre Dupont. English editions followed with translations of the verse.

By far the best-known novel about the Wanderer is Eugene Sue's French work *Le Juif Errant* (*The Wandering Jew*), first serialized in Paris in 1844–45 and published in ten volumes. George Croly's three-volume *Salathiel* (1927, later retitled *Tarry Thou Till I Come*), was an enormously popular earlier novel. (In *Don Juan*, Canto 11, Stanza 57, Byron calls the author "Reverend Roley-Poley.") In Lew Wallace's *Prince of India* (1893), the Wanderer is a wealthy Oriental potentate.

George Macdonald's *Thomas Wingfold, Curate* (1876) introduces the Wandering Jew as an Anglican minister. Having witnessed the Crucifixion, and in constant agony over his sin, Wingfold is powerless to overcome a strange compulsion. Whenever he passes a roadside cross, or even a cross on top of a church, he has an irresistable impulse to climb on the cross, wrap his arms and legs around it, and cling there until he drops to the ground unconscious! He falls in love, but, realizing that his beloved will age and die while he remains young, he tries to kill himself by walking into an active volcano. His beloved follows, but is incinerated by the molten lava. There is a surprisingly happy ending. Jesus appears, forgives the Wanderer, and leads him off to paradise to reunite with the woman who died for him. The novel is not among the best of this Scottish writer's many admired fantasies.

My First Two Thousand Years, by George Sylvester Viereck and Paul Eldridge (1928), purports to be the erotic autobiography of the Wandering Jew. The same two authors, in 1930, wrote *Salome, the Wandering Jewess*, an equally erotic novel covering her two thousand years of lovemaking. The most recent novel about the Wanderer is by German excommunist Stefan Heym, a pseudonym for Hellmuth Flieg. In his *The Wandering Jew*, published in West Germany in 1981 and in a U.S. edition three years later, the Wanderer is a hunchback who tramps the roads with Lucifer as his companion. The fantasy ends with the Second Coming, Armageddon, and the Wanderer's forgiveness.

Sue's famous novel is worth a quick further comment. The Wanderer is Ahasuerus, a cobbler. His sister Herodias, the wife of King Herod, becomes the Wandering Jewess. The siblings are minor characters in a complex plot. Ahasuerus is tall, with a single black eyebrow stretching over both eyes like a Mark of Cain. Seven nails on the soles of his iron boots produce crosses when he walks across snow. Wherever he goes an outbreak of cholera follows. Eventually the two siblings are pardoned and allowed "the happiness of eternal sleep." Sue was a French socialist. His Wanderer is a symbol of exploited labor, Herodias a symbol of exploited women. Indeed, the novel is an angry blast at Catholicism, capitalism, and greed.

The Wandering Jew appears in several recent science fiction novels, notably Walter Miller's *A Canticle for Leibowitz* (1959), and Wilson Tucker's *The Planet King* (1959) where he becomes the last man alive on Earth. At least two movies have dealt with the legend, the most recent a 1948 Italian film starring Vittorio Gassman.

Rafts of poems by British and U.S. authors have retold the legend. The American John Saxe, best known for his verse about the blind men and the elephant, wrote a seventeen-stanza poem about the Wanderer. British poet Caroline Elizabeth Sarah Norton's forgettable "Undying One" runs to more than a hundred pages. Oliver Herford, an American writer of light verse, in "Overheard in a Garden" turns the Wanderer into a traveling salesman peddling a book about himself. "The Wandering Jew" (1920) by Edwin Arlington Robinson is surely the best of such poems by an American writer.

In England, Shelley was the most famous poet to become fascinated by the legend. In his lengthy poem "The Wandering Jew," written or partly written when he was seventeen, the Wanderer is called Paulo. He attempts to conceal a fiery cross on his forehead under a cloth band. In the third Canto, after sixteen centuries of wandering, Paulo recounts the origin of his suffering to Rosa, a woman he loves.

> How can I paint that dreadful day,
> That time of terror and dismay,
> When, for our sins, a Saviour died,
> And the meek Lamb was crucified!
> As dread that day, when, borne along
> To slaughter by the insulting throng,
> Infuriate for Deicide,
> I mocked our Savior, and I cried,
> "Go, go," "Ah! I will go," said he,
> "Where scenes of endless bliss invite;
> To the blest regions of the light
> I go, but thou shalt here remain—
> Thou diest not till I come again."—

The Wandering Jew is also featured in Shelley's short poem "The Wandering Jew's Soliloquy," and in two much longer works, "Hellas" and "Queen Mab." In "Queen Mab," as a ghost whose body casts no shadow, Ahasuerus bitterly denounces God as an evil tyrant. In a lengthy note about this Shelley quotes from a fragment of a German work "whose title I have vainly endeavored to discover. I picked it up, dirty and torn, some years ago. . . ."

In this fragment the Wanderer describes his endless efforts to kill himself. He tries vainly to drown. He leaps into an erupting Mount Etna where he suffers intense heat for ten months before the volcano belches him out. Forest fires fail to consume him. He tries to get killed in wars but arrows, spears, clubs, swords, bullets, mines, and trampling elephants have no effect on him. "The executioner's hand could not strangle me . . , nor would the hungry lion in the circus devour me." Snakes and dragons are powerless to harm him. He calls Nero a "bloodhound" to his face, but the tyrant's tortures cannot kill him.

> Ha! not to be able to die—not to be able to die—not to be permitted to rest after the toils of life—to be doomed to be imprisoned forever in the clay-formed dungeon—to be forever clogged with this worthless body, its load of diseases and infirmities—to be condemned to hold for millenniums that yawning monster Sameness, and Time, that hungry hyena, ever bearing children and ever devouring again her offspring! Ha! not to be permitted to die! Awful avenger in heaven, hast thou in thine army of wrath a punishment more dreadful? then let it thunder upon me; command a hurricane to sweep me down to the foot of Carmel that I there may lie extended; may pant, and writhe, and die!

Scholarly histories of the legend have been published in Germany and elsewhere. In English, Moncure Daniel Conway's *The Wandering Jew* (1881) has become a basic reference. See also his article on the Wanderer in *The Encyclopaedia Britannica*'s ninth edition. Another valuable account is given by Sabine-Baring Gould in his *Curious Myths of the Middle Ages* (second edition, 1867).

The definitive modern history is George K. Anderson's *The Legend of the Wandering Jew*, published by Brown University Press in 1965. A professor of English at Brown,

Anderson made good use of the university's massive collection of literature about the Wanderer. His book's 489 pages contain excellent summaries of European poems, plays, and novels not touched upon here, as well as detailed accounts of the many claimants. The book may tell you more than you care to know about this sad attempt of Christians to avoid admitting that the Galilean carpenter turned preacher did indeed believe he would soon return to Earth in glory, but was mistaken.

ADDENDUM

In Italy the legend of the Wandering Jew took a charming and completely different form. Befana was sweeping her house when the three Wise Men rode by and invited her to go with them to Bethlehem. Befana said she was much too busy. Later regretting her decision, she began wandering about the world under a terrible curse that does not allow her to die. Each year on the eve of Twelfth Night (January 5), a day that commemorates the visit of the Magi, Befana slides down the chimney on her broom to fill shoes and stockings with candy and small toys. She always peers into the faces of the sleeping children, hoping to see the infant Jesus.

Befana's story is told in the following doggerel. I found it in *The Peerless Speaker* (1900) where it is credited to Louise V. Boyd.

"Come forth, come forth, Beffana!"
    She hears her neighbors say,
"Come, up the road to Bethlehem
    The Wise Men pass today!"

So busy was Beffana
    She scarcely turned her head;
Here was the waiting linen,
    The waiting scarlet thread.

Again they cried, "Beffana,
    It is a glorious sight,
Three Kings together journey
    In crowns and garments bright!"

Her people's skillful daughters
    As yet she had excelled.
Beffana saw the spindle,
    Her hand the distaff held;

Her husband's words must praise her,
    Her children's voices bless;
She eateth in her household
    No bread of idleness.

So she made haste to answer,
    "My house is all my care;
No time have I for strangers
    Toward Bethlehem that fare!

"Ere yet the daytime cometh
    I give my household meat:
Mine is the best-clad husband
    That hath an elder's seat.

"And merchants know my girdles
    And my woven tapestry,
The glory of my purple
    And silk most fair to see!"

But now her kinsmen shouted,
    "You know not what you miss!
There may be many pageants,
    Yet none be like to this!

"'Men say the three Kings journey
    A wondrous thing to see,
A babe born of a Virgin
    Foretold by prophecy.

"Oh! come: behold, Beffana!
    For speech may never say
The splendor on their faces,
    The Kings that ride this way!"

Beffana still kept busy,
　But lightly answered then:
"I will look out upon them
　As they come back again!"

But all her friends and kinsmen,
　In wondering delight,
Gazed till the Kings so gently
　Had journeyed out of sight.

That eve Beffana's husband
　Had sorrow in his gaze,
When of her work she told him,
　Anticipating praise.

He did not quite upbraid her,
　But out of ancient lore
He questioned, "Who hath profit
　In laboring evermore?"

And spake of times for mourning
　And times to laugh and sing;
Of times to keep or scatter,
　Of times for everything.

And, sad, Beffana answered:
　"My lord is right, but then
I surely will behold them
　As they come back again."

Alas! alas! Beffana
　Looked out from day to day,
They came no more; God warned them
　To go another way.

And she grew very weary
　Who had so much to do,
And never came the vision
　That might her strength renew.

Beffana dieth never,
　This earth is still her home;
Beffana looketh ever
　For those who never come.

The Wandering Jew and the Second Coming

Many old speaker anthologies contain the following sad poem about the Wandering Jew, translated from the German by Charles Timothy Brooks, a nineteenth-century Unitarian minister, poet, and translator.

The Wandering Jew once said to me,
    I passed through a city in the cool of the year,
A man in the garden plucked fruit from a tree;
    I asked, "How long his this city been here?"
And he answered me, and he plucked away,
"It has always stood where it stands today,
And here it will stand forever and aye."
    Five hundred years rolled by, and then
    I travelled the self-same road again.

No trace of a city there I found;
    A shepherd sat blowing his pipe alone,
His flock went quietly nibbling round,
    I asked, "How long has the city been gone?"
And he answered me, and he piped away,
"The new ones bloom and the old decay,
This is my pasture-ground for aye."
    Five hundred years rolled by, and then
    I travelled the self-same road again.

And I came to a sea, and the waves did roar,
    And a fisherman threw his net out clear,
And when heavy laden he dragged it ashore,
    I asked, "How long has the sea been here?"
And he laughed, and he said, and he laughed away:
"As long as yon billows have tossed their spray,
They've fished and they've fished in the self-same way."
    Five hundred years rolled by, and then
    I travelled the self-same road again.

And I came to a forest, vast and free,
    And a woodman stood in the thicket near;
His axe he laid at the foot of a tree:
    I asked, "How long have the woods been here?"
And he answered, "The woods are a covert for aye;
My ancestors dwelt here alway,
And the trees have been here since creation's day."

Five hundred years rolled by, and then
I travelled the self-same road again.

And I found there a city, and far and near
Resounded the hum of toil and glee,
And I asked, "How long has the city been here,
And where is the pipe, and the wood, and the sea?"
And they answered me, and they went their way,
"Things always have stood as they stand today,
And so they will stand for ever and aye."
I'll wait five hundred years, and then
I'll travel the self-same road again.

LIFE MAGAZINE
AND ASTROLOGY

O n *Life*'s July 1997 cover, spread across pictures of Zodiac signs, are the words "ASTROLOGY RISING: Why So Many of Us Now Believe the Stars Reflect the Soul." The inside cover story is titled "Star Struck: A Journey to the New Frontiers of the Zodiac." The author: someone named Kenneth Miller.

Miller begins by telling how miserable he felt when his first wife fell in love "with a German named Nils." Astrology cheered Miller up. He began buying fashion magazines solely to read their horoscopes. A horoscope-hotline recording by syndicated astrologer Joyce Jillson said the day's lucky color was tangerine. "By chance (or perhaps fate)," Miller writes, "a shirt of that hue lay in my dresser drawer." It was indeed lucky. He put it on for a walk with Julie, a new girlfriend. This led to their marriage.

Intrigued more than ever by stargazing, Miller began to

This essay first appeared in *Skeptical Inquirer* (November/December 1997).

visit top astrologers. He admits that "dozens of scientific studies say it doesn't work," but that didn't faze him. A "physicist-astrologer" named Will Keepin convinced him that astrology had a "scientific basis." The basis? The late Michel Gauquelin's long-discredited claim that eminence in various professions correlates with positions of planets at a person's birth. Miller doesn't tell you that Gauquelin regarded all mainline astrology as rubbish. It's like finding support for a belief that the Earth is flat by reading a book proving its shape is cubical. Miller also became impressed by how physicist David Bohm's emphasis on the interconnectedness of all parts of the universe is also congenial to astrology.

Another astrologer, Karen Helouin, gave Miller a reading in which everything she said hit the mark. She told him he had a need to be of service to people, and that he would have made a good communist. Miller wonders, "Had I told her about my Maoist phase?"

Other astrologers bowl him over with their accuracy. Noel Tyl caught him "rebelling and doing drugs in college." Tyl detected the crumbling of Miller's first marriage and the success of his second one. He is "euphoric" when Tyl predicts "professional ascendency with reward" by April 1998.

Miller looks up Joyce Jillson. He found her horoscope reading "perceptive." A Vedic astrologer impressed him even though he realized that Hindu astrology uses totally different star patterns from Western astrology. I was surprised he did not consult a Chinese stargazer. Ancient Chinese star signs have nothing in common with either Hindu or Western astrology. If one is valid, the other two are bogus.

Miller is now an enthusiastic convert, "overwhelmed by the readings' torrent of insights. . . . I was amazed at the way it shed light on the soul." A visit to Ray Hyman discombobulated him. Hyman, a psychologist and expert in human self-deception, is given two paragraphs to explain how easily believers validate any readings.

"Hyman infuriated me," Miller adds, but the fury was short-lived. Soon he was seeking "treasures in astrology that no fallacy could ever taint. . . . And there were things in those transcripts that Hyman's theory just couldn't explain."

To further bolster his faith, Miller phoned Karen Helouin to give her several dates to analyze. For 1964 she asked, "Did you move?" He did. She guessed correctly it was an unhappy move. For 1987 she asked if he got married. Yes, it was the year of his first wedding. Miller rates Helouin correct five times out of five. His belief in astrology is now unalterable. "Music is beautiful," he ends his article, "to those who hear."

It has been a long time since I read an essay in a reputable magazine by someone as naive, self-involved, and willfully ignorant as Miller. Are there New Age astrology buffs on *Life*'s editorial staff, or was Miller's article no more than a cynical effort to boost circulation? After all, polls show that almost half of Americans believe in astrology, including a recent president and first lady. In either case, *Life*'s editors should be deeply ashamed for their trashy contribution to our nation's dumbing down.

ADDENDUM

The *Skeptical Inquirer* published the following letter by Kenneth Miller and my reply in the November/December 1997 issue.

> The Skeptical Inquirer *invited* Life *magazine managing editor Jay Lovinger to respond to the criticisms in Martin Gardner's News and Comment article. He did not reply, but the article's author, Kenneth Miller, did. His response, too long to print in full, was on* Life *magazine letterhead. Miller identified himself as "staff writer."*
>
> *Here are excerpts:*
>
> "No, there are no horoscope buffs among *Life*'s editors. But a primary function of journalism is to report on

mass culture, and the fact that almost 50 percent of Americans believe in astrology is certainly newsworthy. . . . The article was to be a piece of personal journalism, a well-established genre in which an author discusses his own experiences—even the occasional bout of credulity! —in the hope of illuminating larger truths. . . .

"I did take the scientific critique of astrology seriously, but . . . I also found it fruitful to judge my subject by other criteria. . . .

"I confess that after having my chart read half a dozen times, I briefly became a true believer and that I was rather put out with Ray Hyman for trying to disillusion me. But as I note on page 52, I then checked my transcripts and found that much of what he'd said about selfdeception applied. Nonetheless, his argument didn't strike me as airtight, so I subjected it to an informal empirical trial with Karen Helouin. Her accuracy is almost beside the point. What I found to treasure in astrology was its aesthetically pleasing and psychologically satisfying play of symbols. Music is, indeed, beautiful to those who hear. Mr. Gardner, I submit, has a tin ear."

To which Martin Gardner replies:

With respect to the "beauty" Miller sees in astrology, I confess my ears are made of tin. The solar system is beautiful. The stars are beautiful. Galaxies are awesomely beautiful. But astrology is as ugly as palmistry, phrenology, and reading the future from tea leaves and animal entrails. Its believers have tin brains.

ORAL ROBERTS
ON JIM BAKKER

Recently at an antique mall I picked up a copy of Jim Bakker's *Survival: Unite to Love*, published by New Leaf Press in 1980, several years before Jim's downfall. I bought it because I wanted to read Oral Roberts's introduction.

In light of what happened, the book bristles with laughs. Jim dedicates it "To my wife Tammy who has stood with me through the valley and mountaintop experiences. . . . And to all my faithful partners without whose help there would be no PTL." As we all know, Tammy has divorced Jim, and the PTL partners are still trying to get back some of the loot he stole from them.

Jim is down on greed. "Too many people," he writes, ". . . have made gold their god. . . . Our dividend is not dollars, it's souls. We are placing souls into the bank of Heaven, and that's where our dollars and cents are. That's

This essay first appeared in *Free Inquiry* (summer 1994).

where our profit lies—in souls. Souls, Souls, Souls." A touching photo on page 126 shows Jim weeping as he is "overwhelmed" by "God's goodness." More likely he is overwhelmed by the amount of gold he is putting into earthly banks.

The funniest part of the book is Oral's glowing Introduction. Here are a few gems:

> What drew me to Jim Bakker goes far beyond the splendid work he is doing for God but because he is doing it under public glare two hours a day on LIVE television—warts and all. He stands up and says what he is, what he believes, what he is doing—and when he reaches heart-break hill, he breaks down and weeps like the rest of us mortals.

> Personally, I see a lot of our Master in Jim Bakker—and certainly in Tammy who stands for the Lord even when it is the toughest . . . and when she feels the deepest fears. They both tell me—and say it on the air—they want to be so much MORE like Jesus.

> The clouds are lifting! The dawn is breaking! It won't be long now! The Second Coming of the One Whose we are and Whom we serve is already brushing up against the devil and the opposers of God's anointed. Just one more tiny brush against Satan, and he will be through af-flicting, tormenting, and destroying those for whom Jesus died and rose from the dead.

> This world which wants men and women to be perfect, then tries to crucify them when they don't reach it—and surely would if they did reach perfection—needs to see . . . hear . . . and feel more Jim Bakkers. Yes, because he is becoming a strong Christian and spiritual leader in God's great work here on earth, a true pioneer in the most diffi-cult field of all: being on LIVE television daily in the human arena of grime and blood and suffering of human beings. But also because whatever weaknesses Jim Bakker has he lets [it] hang out for friend and foe alike to see.

Note that last sentence. Read it carefully. Could it be that the Holy Spirit inspired Brother Roberts with the gift of prophecy?

POSTSCRIPT

For more on the saga of Jim and Tammy, see my "Fatherly Advice to Tammy Faye Bakker," reprinted in *On the Wild Side* (Prometheus Books, 1992), and my review of their two recent autobiographies reprinted in this volume.

PSYCHIC SURGERY
IN THE PHILIPPINES

Andy Kaufman, known to television viewers as Latka Gravas, a zany mechanic on the hit show *Taxi*, died of lung cancer on May 16. He was thirty-five. Two months earlier he had gone to the Philippines in a sad, futile effort to be cured by a psychic "surgeon."

For many decades psychic "surgery" has flourished in the Philippines and Brazil. While a patient is fully awake, the surgeon pretends to enter the body with his bare hands, sometimes with a knife as a prop, to remove tissues that are said to cause the ailment. The skin is never punctured. There are no scars, though usually there is plenty of blood.

According to the *Star*, Kaufman was "operated" on twice daily, for a fee of $25 per treatment, by Ramon "Jun" (for Junior) Labo, one of some fifty charlatans who operate in "clinics" in the Philippines. After Kaufman's death the

This essay first appeared in *Discover* (August 1984).

National Enquirer printed three grisly photographs showing Labo's healing session, in which he seemed to pull bloody tissue out of Kaufman's chest. In the first picture Labo's fingers seem to be penetrating the skin. This illusion is produced by bending the fingertips so that the middle knuckles of the fingers press firmly on a patient's body. The tissues and blood, which usually come from animals, are concealed before the operation and produced at the appropriate time by the surgeon, who uses standard magician's sleight of hand to make them appear.

Kaufman's girlfriend, who accompanied him on the trip, was quoted by the *Star* as saying there was no possibility that Labo had used deception, because she stood "not a foot away." "We saw Jun cure a man with an eye problem. He actually removed the eye, and you could see the empty socket. And then he put the eye back in." What he really did is described in detail by surgeon William Nolen in the chapters on psychic surgery in his popular book, *Healing: A Doctor in Search of a Miracle*: the surgeon conceals an animal eye in his closed hand, which he adroitly opens several inches in front of the patient's face after he has pretended to take out an eye. According to the *Enquirer*, Kaufman was as impressed by Labo as was his girlfriend. "The doctors don't know everything," he said in a state of high elation when he returned to California. A few weeks later he was dead in a Los Angeles hospital.

One reason Kaufman went to the Philippines, the *Star* reported, was that he had seen a film about the Filipino surgeons, narrated by that authority on medical science, Burt Lancaster. "We saw the film, and it showed that the cures could work," said Kaufman's girlfriend. What they saw was part of a longer film called *Psychic Phenomena: Exploring the Unknown*, which was aired in 1977 by NBC on Sunday night prime time. The lurid segment on the Filipino surgeons was so effective that promoters of the psychic surgeons have been showing it ever since. NBC replied to the protests of scientists over the show by insisting that it had been produced solely for entertainment. "We can't imagine anybody taking it seriously," one spokesman declared.

Publishers, too, must share the blame for keeping psychic surgeons busy. On the jacket of the book *Psychic Surgery* by Tom Valentine (Henry Regnery, 1973) is this blurb: "The study of Antonio C. Agpaoa, Spiritualist healer of the Philippines, and the astounding facts about successful surgery without instruments, anesthesia or pain." Another offender, *Arigo: Surgeon of the Rusty Knife* by John G. Fuller (Thomas Y. Crowell, 1974), is a fulsome account of the miracles performed by one of the many Brazilian psychic surgeons. Pocket Books has published both in paperback.

In 1975 the Federal Trade Commission ordered four West Coast travel agencies to stop promoting tours of patients to the Philippines, declaring, "Because we are dealing here with desperate consumers with terminal illnesses who want to believe psychic surgery will cure them, no amount of disclosure will suffice to drive home to all the point that psychic surgery is nothing but a total hoax."

Still, the gullible sick keep coming. Business slackened a bit two years ago when the most famous of the Filipino "surgeons," Tony Agpaoa, age forty-three, died of a heart attack. Now the customers are back by the thousands. Every day that Kaufman was at Labo's clinic, as many as a hundred waited in line for his quickie operations.

Perhaps some day the television networks and major publishers, in a fit of moral courage, will realize that when they give invaluable free publicity to medical quacks they are playing with the lives of the innocent and poorly informed. The tragedies occur when the seriously ill, swayed by the glowing testimonials of famous personalities and irresponsible journalists, forgo reputable medical aid until it is too late.

POSTSCRIPT

The following letter ran in *Discover* (October 1984):

> I appreciate Martin Gardner's reluctance to reveal the methods by which legitimate magicians earn a living. However, since he was discussing a man's death and trying to prevent similar suicides-through-ignorance, he might have bent the rules a little.
>
> To a person desperate enough to seek psychic "surgery," the information that the "surgeon" produces tissues and blood by "standard magician's sleight-of-hand" is not a sufficient deterrent. The tissue and blood are in fact produced from a thumb-tip, a flesh-colored, thimble-like device used by magicians for a hundred effects that would not be legitimate to reveal.
>
> But of course all this is redundant, since people gullible enough to fall for psychic "surgery" do not read *Discover.*

> William R. Harwood
> Calgary, Alberta

Of many books praising the paranormal powers of the psychic surgeons one of the most deplorable is *"Wonder" Healers of the Philippines* (1967) by Harold Sherman. This hardcover book reports on three weeks that he and two friends spent in the Philippines to study the surgeons. (On Sherman, see my book *Urantia: The Great Cult Mystery*, Prometheus Books, 1995. Sherman was the author of numerous books on the paranormal, and for many years a devout Urantian before he was expelled from the cult.) Sherman's book purports to be an unbiased account, giving both pros and cons, but actually is a strong defense of the paranormal powers of the psychic surgeons.

Sherman made a second visit to the Philippines in 1967, especially to study the feats of Tony Agpaoa, the most famous and the wealthiest of the Philippine charlatans. His report on this visit can be found in Martin Ebon's

35

The Psychic Reader. Sherman came away still impressed enough to conclude his report:

> Is objective psychic surgery research still possible? In the light of such demonstrations—still inexplicable—mentioned above, it would be most regrettable, even tragic, if charges of fraud and commercialism against any of these "spirit healers" should legislate against continued unbiased and open-minded research. This exploration of the unknown might produce in time, significant discoveries of great possible medical value.

"Dr. Tony," as Agpaoa liked to be called, died in 1982 of undisclosed causes. A few years earlier, when his appendix ruptured, he traveled to the United States to have his appendix removed by legitimate surgeons. For a photograph of Agpaoa "operating," and an accompanying article, see *Time* (October 18, 1968, page 92). Agpaoa dropped out of the third grade, and is said to have been a former magician. He operated on thousands of patients, but never allowed them to keep their removed "organs" to be examined by honest doctors. A good article about him and his associates is William Rice's "A 'Surgeon's' Magic Touch that's too Good to be True," in *Today's Health* (June 1974).

Magician James Randi gave a vivid demonstration of psychic surgery on the *Tonight* show featuring Johnny Carson, operating on a man from the audience who volunteered to be the "patient." Randi got huge laughs at one point when, after removing some bloody goop from the reclining man's bare belly, he said, "That's not supposed to come out," as he pushed the goop back inside.

THE INCREDIBLE
FLIMFLAMS OF
MARGARET ROWEN

B efore Ellen White, the Adventist leader and prophetess, died in 1915 she was careful not to name a successor. However, as years rolled by with no sign of the Second Coming, it was perhaps inevitable that someone would emerge with claims to be the church's new living messenger. That person was Margaret Rowen, a plump, forty-year-old housewife not much larger than a midget. According to her mother she was four feet tall and almost as wide. To this day no one can be sure whether Mrs. Rowen was a self-deluded woman who believed what she preached or a total charlatan. Maybe she was, like Koresh, an evil mixture of both.

Mrs. Rowen's crazy career took off on June 22, 1916, when she claimed she had a vision of soon-coming events. A former Methodist, she had joined the South Side Seventh-day Adventist Church, in Los Angeles, about four

---

This essay first appeared in *Free Inquiry* (spring 1996).

years before her visions began. Little is known about her husband, George W. Rowen, and their three children beyond the fact that they never became Adventists. The Rowens were married in 1899. They lived near Chester and Uplands, Pennsylvania, until they moved to Los Angeles ten years later.

Margaret's visionary trances were allegedly similar to those of Mrs. White. For periods that could last an hour or more her eyes were said to be open and unblinking, her body rigid, and there were no signs of breathing. Tears often flowed down her cheeks. Below are typical descriptions of what she saw as reported in her thirty-two-page pamphlet, "Stirring Messages for This Time." I quote from a series of lectures on Mrs. Rowen delivered some time between 1972 and 1974 to a General Conference of Adventists in Takoma Park, Maryland. The speaker was Arthur L. White, a grandson of Ellen White. His unpublished lectures, which came to me by way of the Adventist historian Richard Schwarz, are a major source for this article.

> I stood with the angel Gabriel in the paradise home of Adam and Eve and saw them as they came from the hand of their Creator. Majestic in stature they were, and beautiful, perfect in every way, and wearing a soft covering of light. Among the beasts of the garden was one more beautiful and intelligent than any of the others. This little four-footed animal stood upright, front feet were like hands being used to convey food and drink to the mouth. Its yellow and black striped body was covered with soft silky hair and shone like gold. Two large gauzy wings which sparkled with a silver light enabled the pretty creature to fly high among the lofty branches of the trees. This creature, the serpent, was ever Eve's pet. It was her daily habit to amuse herself with this beautiful animal.
>
> Were I drawing only beautiful pictures, I should omit what follows, but will give as clearly as possible the scenes as they passed before me. Satan shows the effect of his long existence apart from God. His skin is almost brown and hangs in loose folds from the lower face, while

under the chin, the bagginess falls like a pointed beard of flesh. The perceptions are over-developed, giving the most unnatural appearance to the eyebrows which lift in great arches above the immense eyes, beautiful at one time. There is much loose flesh above and below the eyes which sometimes bulge, sometimes sink back into the head, and at still other times, roll uncontrollably from side to side. The brow is high, receding to the hairline. The hair is of a dirty slate color, coarse and bristly. Patches of this same bristly hair appear on different parts of the body which is bent, twisted, and almost fleshless. The most unsightly is the mouth which lifts high at the corners, the upper lip drawing far down at the center. The frightful teeth growing far out from the gums lend much to the hideousness of the grin that he ever seems to wear. This is the monster that uses his influence to trouble Pilate and his company at every turn.

Margaret's description of Satan is similar to what Mrs. White saw in a vision reported in her *Spirit of Prophecy* (vol. 1, p. 48). Although Mrs. Rowen's visions usually harmonized with those of Mrs. White, there were occasional departures from Adventist beliefs. For example, Mrs. White taught that Jesus and Michael were one and the same, but Margaret added the heresy that Jesus was not part of an eternal Trinity. He was an archangel, created by God and later adopted as God's only son. Not until then did he become a member of the Trinity. Margaret also preached that during the Millennium, while the saved were enjoying heaven, Satan and his fallen angels, along with a resurrected Pilate and all the false prophets, would be allowed to wander freely over a desolated Earth. Her most peculiar revelation was that the 144,000 redeemed, mentioned in the Apocalypse, would be a subset of the saved and gathered solely from the United States.

A small number of Adventist elders became convinced that Mrs. Rowen's visions were genuine. Arthur White singles out three: P. W. Province, of Oregon, and two Los Angeles ministers, J. F. Blunt and F. I. Richardson. Blunt published a forty-page booklet in 1918 titled *The Rowen Pamphlet* that vigorously defended Mrs. Rowen as the church's new messenger. Two other prominent Adventists who supported Mrs. Rowen were Elder Julian M. Tvedt, of Coffeyville, Kansas, and his friend Elder Matthew Larson, pastor of a church in Arkansas, Kansas.

Careful investigation by the Adventist church finally concluded that although Margaret's visions were supernatural in origin, they came from Satan, not from God or the Holy Spirit. Mrs. Rowen and all the elders who followed her were disfellowshipped (excommunicated) in 1919 along with former Iowans Dr. Bert E. Fullmer, a private physician, and his wife, Jessien. The Fullmers had become Mrs. Rowen's most devoted disciples.

It is impossible now to know how many Adventists joined Mrs. Rowen's band of apostates. A reasonable guess is about a thousand, most of them in California, but also scattered throughout the United States and abroad. Margaret was tireless in preaching here and there after her excommunication, issuing a raft of handbills and pamphlets about her visions and prophecies. Many of her handbills were dropped from airplanes. She and her husband lived in a duplex apartment in Los Angeles. The adjacent apartment was occupied by Dr. Fullmer and his wife, adventists from Iowa. Dr. Fullmer had been an elder until he resigned the ministry in 1908. Fulmer became Mrs. Rowen's publisher, and Mrs. Fullmer served as secretary and treasurer of what was called The Seventh-day Adventist Reform Church. Members in Los Angeles met regularly for Saturday services at Rhodes Hall, at the corner of 55th Street and Moneta Avenue. For four years. Dr. Fullmer published Margaret's pamphlets and leaflets, and edited the movement's periodical, the *Reform Advocate and Prayer-Band Appeal*. Copies of this journal are now exceedingly rare collector's items.

Margaret promoted her reform movement with outrageous deceptions. She announced that she had a vision informing her that Ellen White had signed a document on August 10, 1911, naming her (Mrs. Rowen) as her successor, and that this document was hidden in a vault at Elmshaven, an Adventist medical center in St. Helena, California, where Mrs. White had spent the last fifteen years of her life. Margaret confided to Dr. Fullmer that she had visited Elmshaven and secretly removed from the files this important document which the church was concealing. She regretted having done this. The document belonged back in the vault. She begged Fullmer to visit Elmshaven and surreptitiously return the document to the files.

The document, typed and apparently signed by Mrs. White, read as follows:

St. Helena, California
August 10, 1911

In the night season the spirit of the Lord came upon me and I was shown the great falling away among the remnant. Many times the written testimony was read from the sacred desk, and the reader himself failed to apply it to his own heart and profit by it. Rather than shape their lives by the words from heaven, they would permit that which condemned them to be wiped from the printed page. But I saw that God would keep his promises to Israel and would have a people—a remnant unspotted from the world. To accomplish this I saw it was necessary to call for a reform in the church of God. I saw that the spirit of God moved upon a few to seek for a purification of heart. This work would rise and fall again and again; the allurements of the world seemed more than they could withstand. I then saw that just a little way in the future after my labors were finished, that God would call one to give the cry to the church of god, "Repent. Repent. Lift up the standard. Purify yourselves for the coming of your King." I then saw little companies formed to call on God for help. I saw these earnest praying ones much annoyed by their brethren, Criticized and condemned, the messenger of God suffered much humiliation and condemnation, but she longed to see the church made white and clean, ready for the Saviour's coming. Through all of her trials she was true to her calling. In her heart was the love that the Master would have for the brethren. I saw that many were shaken out because of the straight testimony, but the little praying companies increased until the Church of God was honeycombed with earnest, praying people. I saw that many of the leaders refused to accept the messenger. I saw that the one sent of God was one of limited education, small in stature, and would sign the messages Margaret W. Rowen.

[Signed] Ellen G. White

The Fullmers, completely under the spell of Mrs. Rowen, believed everything she said. Together they drove to St. Helena. While a secretary went out of the room to get a lamp—the vault was dark—Dr. Fullmer slipped the document into the files.

At the urging of Mrs. Rowen, the files were searched, and the document was discovered in the vault on December 17, 1919. It was immediately recognized as a crude forgery, and before the year ended Margaret and both Fullmers were excommunicated.

Mrs. Rowen never admitted that she had forged the document. In 1924, she produced a letter purporting to have been sent by Ellen White's son William Clarence to Mrs. Rowen's follower, the former Elder Richardson. In this phony letter, which she had also forged, White confessed that he had destroyed his mother's original letter, replacing it with a spurious copy so easily seen to be a forgery that it would totally discredit Mrs. Rowen. Margaret then produced a much better forgery of the original document, which she claimed had not been destroyed after all!

Shortly before he died in 1927, Dr. Fullmer made a full confession about his role in slipping the first document into the files at Elmshaven. To ease his conscience, he said, he had vowed to himself that if anyone ever asked him about the matter he would tell the truth. To his amazement no one ever asked.

Adventist leaders had long noticed that Margaret, like so many cult leaders, enjoyed a lavish lifestyle. She wore expensive clothes and drove high-priced cars. Where was the money coming from? Actually, it came from the tithes of her deluded followers, which she secretly diverted to her own bank accounts. To explain her lifestyle she fabricated the following story.

She had always believed, she said, that she was the daughter of Alfred and Mathilda Wright, then living at 339 East 38th Street, Los Angeles. To her astonishment, a vision revealed that her true mother was one Mary Gillette. Mary had been seduced by her father before he married Mathilda. In the vision she saw herself as an illegitimate baby being put in a basket and carried by her father to his house where he told his wife he had found the child abandoned on a wharf. The Wrights loved the child and later adopted her.

Margaret's true mother, Mary Gillette, was now married to a wealthy Philadelphia businessman named Harold Mills. After learning his address from another vision, Margaret contacted Mrs. Mills by mail and quickly convinced her that she was indeed her daughter. Mrs. Rowen showed letters from Mrs. Mills to her "precious one" in which she spoke of longing to see and embrace her "darling" daughter. Margaret drove to Philadelphia where she said she spent five weeks with the Mills family. She produced a photograph of Mrs. Mills that showed a remarkable resemblance to herself. Mrs. Mills was so overjoyed to discover that Margaret was her daughter that she began giving Margaret large sums of money and a regular allowance. She promised Margaret she would be the heiress of a huge estate.

Mrs. Rowen told her followers that the Mills family was coming to Los Angeles to visit her on December 26, 1918. Unfortunately, Mr. Mills came down with the flu, so the trip was canceled. He died, Mrs. Rowen said, on Christmas Day.

His widow was too distressed to make the trip to Los Angeles, but she continued to send money.

It was discovered much later by Adventist leaders that this story was pure fabrication! There never was a Mills family or a Mary Gillette. It was nothing more than Margaret's wild cover story to account for the source of her funds. True, she did make a trip to Philadelphia, though not to spend with the imaginary Mrs. Mills. Instead, she had stayed for five weeks with her younger sister, Mrs. Sue Henley, in Chester, south of Philadelphia. The photo of Mrs. Mills had been taken in 1917 by C. F. Havercamp, a commercial photographer in Chester. Mrs. Rowen had one picture taken of herself. Then she used make-up to make her look like an older Mrs. Mills for a second picture, so naturally there were strong facial resemblances in the two photographs. She asked Havercamp to print the two pictures side by side on a single card.

In 1920, Adventist investigators found a telegram that Mrs. Rowen had sent to Havercamp. In it she ordered him to send her all prints and negatives, threatening him with legal action if he ever revealed what he had done. This information, by the way, came from Sue, who had no respect for her older sister.

Neither had Mrs. Wright. In 1920 she signed a sworn deposition which read:

> Any statement to the effect that Mrs. Margaret W. Rowen is not my natural daughter, but a foundling adopted by me by any process, legal or otherwise, is unqualifiedly false, as she is bone of my bone and flesh of my flesh.

Similar statements were obtained from John and James Plummer, Margaret's two half-brothers by an earlier marriage of Mrs. Wright. Mrs. Wright told investigators, "I don't know why Margaret does this. I think she is trying to get money out of it some way. I wish she wasn't my daughter." John Plummer said he was more sure of Margaret's birth than of his own because he knew nothing about his own birth whereas he himself had gone for the doctor when Margaret was born. He agreed with his mother that Margaret was under the control of a "hypnotic or occult power" and that her family was "disgusted and humiliated by her claims."

In 1920 the Adventist Church's General Conference Committee published a forty-eight-page pamphlet about Mrs. Rowen titled "Claims Disproved." It was followed in 1923 with a sequel, "Further Statement About Claims Disproved."

Mrs. Rowen never gave up trying to convince her followers that the Mills family existed and was the source of her funds. More and more forged documents were produced. She even hired an actress to impersonate Mrs. Mills, introducing her to the Fullmers and other followers. At another time she persuaded a friend, Mrs. Helen Despat, to wear a black veil and accompany her to church posing as Mrs. Mills.

In 1920, Margaret printed a handbill telling how she had been injured while speaking in a hall in Spokane, Washington. She claimed that a Catholic man in the audience, angered by her attacks on the papacy, had pulled her off the platform and broken her arm. She said the Spokane police had falsely arrested her along with her assailant. A check with Spokane's police revealed that the entire story, as the chief of police put it in a letter, was "a fake, pure and simple." His department knew nothing about a Margaret Rowen.

How can one explain Margaret's never-ending compulsion to make crude forgeries and to concoct easily detected whoppers?

It is difficult to believe she was told to do this while in a trance. Even her visions are suspect because the only document describing them—the open eyes, the rigid body, the absence of breathing—proved to be a forged certificate signed by Dr. E. C. Cavenaugh, of Spruce Street, Philadelphia. There was no physician by that name at the time in the Philadelphia area.

*   *   *

On the other hand, in 1923, Mrs. Rowen made a mistake so stupid that one is inclined to suspect she may really have experienced visions and believed them. Like Baptist farmer William Miller, whose failed prophecies about the Second Coming were the origin of Seventh-day Adventism, she set an exact date for the arrival of Jesus. It was to occur on February 6, 1925. Probation for the unsaved, she said, would close a year earlier on February 6, 1924. Handbills were printed announcing the dates.

The first sign of Jesus' return, Margaret said, would be a small black cloud that would appear in the east after midnight on February 6, followed by a great search light from heaven. The black cloud came from one of Mrs. White's visions. In her *Early Writings* (page 11) she wrote: "Soon our eyes were drawn to the east, for a small black cloud had appeared, about as large as a man's hand, which we all knew was the sign of the Son of Man." In the vision she saw Jesus seated on the cloud.

The cloud, according to Mrs. Rowen, would signify the start of Christ's seven-day journey from heaven to earth, accompanied by a vast host of angels. Along the way they would stop off to visit inhabited planets and tell the residents the good news. On the return trip, which also would take seven days, the angels and the redeemed souls from Earth would spend a sabbath of rest on one of the planets. Those saved would include 144,000 from different parts of the United States. They would be teleported from their homes to California before being lifted into space to join Jesus and the angels. Terrible plagues, earthquakes, and other disasters would then reduce the Earth to rubble.

Throughout the United States, the Rowenites sold their homes and worldly goods to await what is now called the "rapture." Stories about them appeared in newspapers throughout the land. A *San Francisco Chronicle* report was headlined: "Faithful Await Doom of World—Wicked

Promised Great Free Show." The editor of a Pennsylvania paper wrote: "Will all those who have not paid for their subscription please call at the office tonight before midnight to save us the task of scurrying all over hell to try to collect past due accounts."

Little is known about how Mrs. Rowen reacted on February 6, 1925, when nothing happened. She surfaced later that year to say that it was taking Jesus longer to make the journey to Earth than she had expected, but that he would be arriving very soon and terrible things would happen to the world after the faithful, including herself, were raptured.

In 1926, the long faithful Dr. Fullmer finally began having doubts about Mrs. Rowen's honesty. Surely her failed prediction of the date of the Second Advent must have played a role in his disenchantment. In his periodical, he published uncontrovertible evidence that Margaret had twice forged the document supposedly signed by Ellen White, as well as the letter in which Elder W. C. White confessed that he himself had forged the first document. Fullmer reported on his investigations of Margaret's funds. They proved that she had been regularly stealing tithe money from her followers to pay for her costly wardrobe and the five expensive cars she had bought over the years. Fullmer apologized to his readers for the role he had played in duping them and expressed his belief that Mrs. Rowen was under the influence of Satan. He made plans to take her to court on a charge of stealing tens of thousands of dollars from the faithful.

It is hard to believe, but Mrs. Rowen, now thoroughly discredited and desperately in need of money, decided that the only way to stop Fullmer's attacks on her was to kill him! On Sunday, February 27, 1927, Dr. Fullmer had a midnight phone call about a sick child in Cabin 11 of a motel near Lankershim. He hurried there with his little black bag. In the cabin awaiting him were two loyal aides of Mrs. Rowen: Dr. J. Franklin Balzer and Miss Wade, a nurse. Dr.

Balzer's wife had recently divorced him over his affair with Miss Wade, and was suing him for back alimony.

Balzer knocked Fullmer unconscious by batting him on the head with a lead pipe. As he regained consciousness, an effort was made to inject poison into his arm, but during the scuffle the hypodermic needles broke, saving his life. Tenants in nearby cabins phoned police. When they got there, Balzer and Wade had fled. Mrs. Rowen had been present earlier, but left before Fullmer arrived. Police found in the cabin a shovel, a rope, a blanket, and a large piece of canvas to be used for burying Fullmer's body in a nearby desert.

Mrs. Rowen and her two accomplices were arrested. All three were sentenced in 1927 to from one to ten years in San Quentin prison, which then accepted women. The charge was assault with a deadly weapon with intent to do great bodily injury.

Fullmer went ahead with plans to sue Mrs. Rowen as soon as she was released from prison, but, before 1927 ended, he died of a heart attack. His widow rejoined the Adventist church.

Mrs. Rowen was freed on parole after about a year of good behavior. She at once broke parole and vanished. In 1931 she turned up in Miami, Florida, where she was holding prayer meetings and travelling in a Ford car with a Mr. J. J. Hartman. They told the clerk of a motel where they had bunked that they were missionaries from San Diego.

Margaret is believed to have died in the late 1940s, though where and when she died, and under what circumstances, remains a mystery. If any reader has information about her death, please let me know. I would also be grateful if someone in Los Angeles would copy for me, from local papers available in libraries, reports on the attempted murder of Dr. Fullmer, and the arrest and sentencing of Mrs. Rowen and her two partners in crime. I will reimburse copying expenses and pay extra for labor and time. Of

Woman Cult Leader Held On
$2,500 Bail In Murder Plot

LOS ANGELES, Calif., March 9 *1927*—Mrs. Margaret W. Rowan, religious leader, who at one time called upon members of her following to prepare for the end of the world today was held in $2,500 bail for preliminary hearing tomorrow on a charge of conspiracy to commit murder.

In arraignment proceedings she was accused of having plotted to slay Dr. Burt E. Fullmer, leader of an opposing cult. Fullmer charged that he was lured to a tourist camp near here and drugged and given a beating. He said the attack followed his threats to expose the activities of Mrs. Rowan and her followers.

Dr. Jacob F. Balzer and Miss Mary A. Wade, members of Mrs. Rowan's cult, have been named with her in the conspiracy complaint, after being arrested at the scene of the asserted attempt to kill Fullmer. They admitted the attack, implicating the leader of their cult as the instigator.

Mrs. Rowan disappeared for more than a week following the attack on Fullmer. Her followers told authorities that she had been "called over the mountain tops." When Mrs. Rowan reappeared late yesterday and surrendered to the police, she denied that she participated in the attack on Fullmer. Instead, she said she was attacked after complying with a telephoned request to appear at the tourist camp. She escaped and made her way to Phoenix, Arizona, by begging rides from motorists after her own car had broken down near San Juan Capistrano.

Efforts of the cult leader to drown herself when she "became tired of everything," while on her way to Arizona, were frustrated, she said, when the waves at San Juan Capistrano beach washed her out of the sea.

Mrs. Rowan hobbled into police headquarters on crutches. Her son, B. A. Rowan of San Bernardino, at whose home she appeared on her return from Arizona, had persuaded his mother to surrender to the po-

ABOVE—Mrs. Margaret W. Rowan, who after disappearance of week during which her followers told authorities she had been "called over the mountain tops," faces charge of conspiracy to to commit murder. BELOW—Dr. Burt Fullmer, rival cult leader, her accuser.

The only known photo of Mrs. Rowen.

course I would also be delighted to hear from any of Mrs. Rowen's descendants.

Only one photograph of Mrs. Rowen is known to exist. It is reproduced on page 78 of an article about her in *Notes and Papers Concerning Ellen G. White and the Spirit of Prophecy*, revised edition 1974, issued by the White Estate. The Rowenite movement is also covered briefly in the *Seventh-day Adventist Encyclopedia* (1966).

THE SAD SAGA OF
DR. BERT FULLMER

I n part one, I told the bizarre story of how Mrs. Margaret W. Rowen, of Los Angeles, tried to establish a break-away movement called the Reform Seventh-day Adventist Church. She claimed to be God's chosen successor to Ellen White, the inspired prophetess who, with her husband, James White, had founded the Seventh-day Adventist Church.

Mrs. Rowen seemed to experience trance visions similar to those of Mrs. White. God, the Holy Spirit, the angel Gabriel, and sometimes Jesus himself would use her as a channel for new revelations. The most startling of these disclosures was that Jesus would return to Earth in glory on February 6, 1925. When this momentous event failed to take place, the reform movement naturally collapsed.

Mrs. Rowen's most dedicated disciple was Dr. Bert E. Fullmer, a Los Angeles physician and former Adventist

This essay first appeared in *Free Inquiry* (fall 1996).

elder. For many years Fullmer and his wife, Jessie, were totally persuaded that Mrs. Rowen was what she claimed to be, a messenger God had chosen to energize a lukewarm Adventist denomination and to announce the day of Jesus' Second Coming. As official publisher of the Reform Church, Fullmer printed Mrs. Rowen's pamphlets and leaflets. They were widely distributed throughout the nation, some even dropped from airplanes. Fullmer also edited and published *The Reform Advocate and Prayer-Band Appeal*, a twelve-page, usually bimonthly, periodical that flourished from its first issue in August 1922 to its final issue in August 1926. It often carried articles by Mrs. Rowen. Thanks to generous assistance from Adventist libraries, I was able to obtain a complete run (except for two issues) of this peculiar journal.

Reading through the four years of *The Reform Advocate*, one is astounded by the gullibility of the Fullmers in taking seriously Mrs. Rowen's preposterous claims. Although strong supporters of the Reform movement, they never ceased to consider themselves orthodox Seventh-day Adventists who accepted without questions the genuineness of Mrs. White's visions.

Early issues of *The Reform Advocate* focused on defending Margaret's most significant departure from Adventist doctrine. The church taught that, at the time of Christ's return, the 144,000 saints cited in the Book of Revelation (chapters 7 and 14) would consist of all the redeemed who steadfastly refused to accept the Mark of the Beast, a mark taken by Mrs. White to symbolize the Roman Catholic church's unwarranted change of the Sabbath from Saturday to Sunday. This was a plausible view in Mrs. White's day, Fullmer and Mrs. Rowen argued, because at that time there was only a small number of Adventists around the world. But by 1922, Fullmer pointed out, the number of Adventists in the United States alone was far more than 144,000. Surely, he insisted, the number to be raptured when Jesus returns will be much greater than 144,000.

Who, then, are the 144,000? Fullmer and Mrs. Rowen agreed with Ellen White that the two-horned beast of the Apocalypse signified the United States. During the years of the great Tribulation, Satan, incarnated as the Antichrist, would do everything in his power to thwart the remnant church and persecute its members. Those in the United States who remained steadfastly faithful would constitute a special subset of the saved. When Christ returns, before the tribulation begins, this little band of saints would be transported to a spot in California, Mrs. Rowen taught, where they would be caught up in the air and carried to heaven by Jesus and a host of angels led by Gabriel. Fullmer printed and offered on sale for ten cents an eighty-one-page pamphlet titled "The One Hundred Forty-Four Thousand."[1]

It was not until 1923 that Dr. Fullmer announced in his periodical that Margaret Rowen had had visions in which three important dates were revealed to her. On July 23, 1923, Jesus had begun his work of judgment on everyone living. Probation for the unsaved would close on February 6, 1924. Exactly one year later, at midnight on February 6, 1925, Christ would return to Earth. After leaving with the 144,000 saints, the Earth would be decimated by terrible earthquakes and plagues. Indeed, some of the plagues, Margaret announced in 1924, had already started. The Earth would be turned into a dark, desolate region where Satan, along with a resurrected Pilate and other evil persons of the past, would roam about in the darkness awaiting their eventual annihilation in a Lake of Fire. It would be the end of the world as we know it.

Here is how Mrs. Rowen phrased it in *The Reform Advocate* (July 1924, page 6):

> We are living in the last year of earth's history. We are living in the time of the falling of the plagues. We are living in the time when God's saints will not only go through the perfecting process of the time of trouble; but we are living in the time when they will unite, with the power of God upon them, in proclaiming the Sabbath

> more fully, and calling the saved of every denomination to
> come out of Babylon, that they receive not of her plagues.
> [Rev. 18:1–14]

Poor, naive Doc Fullmer bought it all. Margaret's prophecy of doom was trumpeted by Rowenites around the globe. The Adventist church, however, was understandably appalled. As we learned in the previous chapter, it excommunicated Mrs. Rowen, Dr. Fullmer and his wife, and numerous church elders who had been caught up in the Rowenite hysteria.

As Adventist leaders repeatedly pointed out, the Bible declares that no one, not even Jesus himself, knows the time of the Second Advent. How did Fullmer respond? This was true, he reasoned, when the Gospels were written, but no longer true today. Elaborate preparations obviously must be made in heaven for such a stupendous event as Jesus and his angels traveling all the way to Earth on a cloud. Was it not unthinkable, Fullmer asked, that the Lord would not know the date on which this long trip would take place? He reminded the church that back in 1844, when William Miller had set a date for the Second Coming, his followers who later became Seventh-day Adventists accepted Miller's date without any qualms.

In the December 1923 issue of his journal, two months before the day Jesus was expected to appear in the east, Fullmer wrote that during his eight years of knowing Mrs. Rowen he had never observed the slightest hint of deliberate deceit on her part. If Jesus fails to arrive on February 6, Fullmer promised to openly admit that he and his wife, and Margaret, had been cruelly bamboozled by Satan. If her vision of the end of the world did not come from Gabriel, as Margaret claimed, it could come only from the arch-fiend.

So strong is the power of rationalization that the good-but-stupid doctor was incapable of keeping his view. When February 6 passed, with no sign of the cloud, Fullmer could not believe that Margaret's prophecy was totally false. In the February–March 1925 issue of *The Reform Advocate*, he reported that Mrs. Rowen had taken the non-appearance of the Savior very hard. "No person has suffered more than she in this experience," he wrote. "She does not expect that any word from her will carry weight with many, under present conditions, and earnestly hopes there will never again be laid upon her another such burden."

In his July issue Fullmer wrote that Margaret was "very much broken in health." Her reaction to the failure of her prophecy suggests that her visions of the end may really have come to her in actual trances. It is hard to imagine that she could have been so foolish as to announce a date for the world's end without believing this had been revealed to her from on high.

For several more issues Fullmer struggled mightily to find a way out of his deepening perplexity. He still believed that February 6 was the date given to Margaret by Gabriel. After all, he reminded his readers, Miller's 1844 date was later taken by Adventists to be correct, but the event that took place that year was not the date of the Lord's return. It was the date Jesus entered a Sanctuary to begin judging the living. The date was right. Only the event was wrong.

Could something like this apply to Mrs. Rowen's seemingly failed date? It was possible, Fullmer reasoned, that they had not taken into account the time it would take for Jesus, riding on the cloud with his angels, to get from heaven to Earth. Could February 6 be the day Jesus mounted the cloud? "We are in disgrace," Fullmer admitted. He and the flock of Rowenites were being "held up to ridicule and scorn" by newspapers throughout the land. Believers in Mrs. Rowen turned on her and on him with vicious anger. Even their lives had been threatened. The mainline Adventists were crowing loudly over how right they had been all along in branding Mrs. Rowen a false prophet and excommunicating her.

It took Dr. Fullmer more than a year to admit that Margaret's visions were Satan inspired. He struggled during his long period of confusion to hope and believe that light eventually would come from above to explain Mrs. Rowen's seeming blunder. He still was certain that Margaret was totally sincere. He wrote that he had not lost his faith in God or in the soon return of the Master, or in any of the basic doctrines of Seventh-day Adventism. He said he intended to continue publishing his periodical, awaiting the time when God would provide an explanation of what disenchanted Rowenites were now calling the "Crash."

In April 1925 Fullmer reported that he was being accused of offering a "poor alibi" for the failed prediction. In his May–June issue he still clung to the notion that February 6 was the day Christ began his long journey, not the day of his arrival on Earth. However, he was honest enough to print letters coming to him from furious Rowenites and Adventists. Here are two of them:

> Dear Old Friend:
>
> We are sorry for you, and now since you staked everything on the prediction of Mrs. R. concerning the end of the world coming Feb. 6th, 1925; and now when her prediction proves to be a falsehood, you still refuse to say that she was a false prophet, we are ashamed that an old friend of ours should take such a position. . . .
>
> You had your answer Feb. 6th, 1925. Let that settle it as you publicly proclaimed about one year before that you would! When you do *not* let that settle it, would you like to know how some look upon your judgment? "O wad some power the giftie gie us, to see oursels as ithers see us."
>
> I have received a letter from one of your oldtime friends in which is a paragraph, I think I will quote to you that you may have a fair chance to know some of those who were your real friends in years gone by feel about your present attitude. . . .
>
> Here it is:
>
> "I see Fullmer is still publishing his untruth and slush. Poor deceived man and still they push on. I was going to write to him but I did not. I feel so ashamed of him. I never tell any one I know him any more. I believe that God is ashamed of him."

Now, ———, this is not very complimentary to your judgment. . . .

With sincere wishes for your deliverance.

A Conference Pres.

Dear Doctor:

You are most interesting in your reasoning powers. Your past arguments in favor of the date were upheld by your far-fetched arguments, and now you have still more. It is interesting as a mental exercise to follow you, . . .

If you would apply to yourself what justly belongs there, you would be better off, . . . It seems a great pity to me that you, with your magnificent power, and pleasing personality should be kept out of proclaiming the truth to the world. Seven long years you have been pecking away at faults in the church, both real and imagined.

You seem above all common missionary work, having apparently a God-given charge of being a "corrector of heretics."

. . . Why did you yield the first step, away back there in 1908, in giving up the ministry?

You have made the name Adventist a laughing-stock over the globe. Your recent publication reveals you to be one of "those presumptuous spirits which will not accept rebuke."

How a man can be so self-deceived is more than I can see. The more difficult is the matter, because your life is apparently (and I believe actually) clean. If you were not so, the task would be simpler.

Think of the money you have spent yourselves, and caused the S.D.A. to spend refuting you, which Satan has kept out of the cause of truth. . . . But, "presumptuous spirit" you "will not accept rebuke."

Do not understand me, Doctor, to be against a reform. God knows we need it. Our people as a whole were never more worldly than they are now. . . .

God bless you, Doctor, you were the ideal of my youthful heart when I decided to become a minister. I wanted to be like Elder Fullmer, but you have been a poor example in your mad career.

Very sincerely, A Minister

Fullmer's March 1926 issue was the most important, the most sensational, and the saddest of all the issues. For the first time he openly accused Mrs. Rowen of deliberate fraud. The periodical's first page was headlined "Unveiling the Mystery. Innocence Vindicated. Guilt Proven. Apologies Offered."

"To some it will seem as the death of their next of kin," the doctor wrote, "because of the dissipation of their fondest hopes." To others "who have looked upon the matter as a joke, they may be inclined to laugh at the outcome." Fullmer then followed with a detailed confession of the unwitting but deceptive role he played in connection with what he now knew to be Margaret's forgery of the letter in which Mrs. White named Margaret as her successor. He also provided evidence of her forgery of other letters. "With disappointment, sadness, and regret," he said, "it is my duty to face Mrs. Rowen with the evidences upon which I am compelled to make the charge."

Fullmer spoke of his wife's devotion to Mrs. Rowen as a trust that "has amounted to thralldom." Addressing his wife directly:

"Your early contact led you to a faith in her [Margaret] and her experiences which knew neither bounds nor respite. Believing that this was of God, your obedience was absolute, developing into a fear, either to disobey or question." He added that his wife, Jessie, out of love and devotion for him, has offered to take all the blame for their years of being duped. The doctor refused to do this. "I demand my share," he added in italics.

To Mrs. Rowen he said: "Sister Rowen, I am doing the saddest task of my life as I write these lines. . . . Yet I am writing as a friend which I have promised to try to be, even in such a crisis as this." He urged her to make a full confession of her deceptions as a first step toward her freedom from the satanic powers that had been her undoing.

"There are many who have to bear this struggle, and suffer in heart over that which is here written," but the

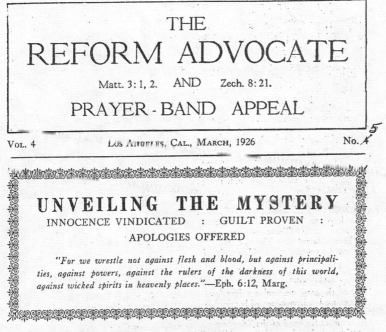

THE
REFORM ADVOCATE
Matt. 3:1, 2.   AND   Zech. 8:21.
PRAYER - BAND   APPEAL

Vol. 4   Los Angeles, Cal., March, 1926   No. 5

UNVEILING THE MYSTERY
INNOCENCE VINDICATED   :   GUILT PROVEN   :
· APOLOGIES OFFERED

"For we wrestle not against flesh and blood, but against principali-ties, against powers, against the rulers of the darkness of this world, against wicked spirits in heavenly places."—Eph. 6:12, Marg.

THAT which appears in this article will come as a distinct surprise to all who read it. The reactions provoked by it will vary with the emotions of the individuals.

Some will hail it with thanksgiving, because of their anxieties regarding the subject in hand.

Some will receive it with relief, because of the vindication it will prove to them.

To some it will be as the death of their next of kin, because of the dissipation of their fondest hopes.

Others will find themselves involved in a maze of perplexities, either of doubt or fear.

A few will be overwhelmed with chagrin or shame, because their actions will have been shown ill-timed and un-wise.

The settling of the question with *certainty* will bring a sense of relief to those who have been heart-burdened as to whether they were being led astray or rejecting truth.

The ever-present wiseacre will appear and affirm, "I knew it all the time," but untruthfully; for a conviction or impression falls far short of a conviction formed by evidence and proof.

If there are any who have looked upon the matter as a joke, they may be inclined to laugh at the outcome, and enjoy the point.

Unstinted blame may be attached to some, and the motives underlying their actions challenged; and these must of necessity meet this attitude on the part of those so moved.

On the other hand, more than one whose motives or actions, or both, have been challenged will find themselves cleared of all imputation of either known or suspected deceit or wrong-doing.

That unsuspecting persons have been innocent instruments in furthering unknown design, will also appear in several instances.

Lastly, the truth of the opening text will be demonstrated, that our battle is not, and has not been, with matters of human design; but a conflict with a power whose reality and subtlety can be questioned only by challenging the truthfulness of the Bible.

370

The first page of Dr. Fullmer's periodical in which he reveals for the first time his discovery that Mrs. Rowen is a false prophet and a charlatan.

"greatest struggle of all wages in the bosom" of Margaret. Having now made a full confession of the role he played with respect to her forgery of a letter signed by Ellen White, he now felt he had finally laid down his burden. "Across the closing chapter of the story of Margaret W. Rowen, there stands written the final word, FINIS."

The dark cloud bearing Jesus was still nowhere to be seen, but even darker clouds were still gathering over Fullmer's head. In June–August, 1926, he disclosed that Mrs. Rowen, along with several companions, was visiting Adventist churches here and there to report on new visions and to try to obtain much-needed money. He warned church members to "Keep your eyes and ears open, and your hands upon your pocketbooks, and stay at home."

Mrs. Rowen and her friends had started a new periodical called *The Pathway*. "Whither does it lead?" Fullmer wants to know. Its first issue reported a stupendous miracle. On the night of May 31, 1926, at about 12:20 A.M., Mrs. Rowen was said to have died "in the presence of family, physicians, and nurses. . . . Her body was made ready for the undertakers and all funeral arrangements were completed." A short time later, *The Pathway* continued, Mrs. Rowen was resurrected!

Fullmer refused to believe this. He demanded to see a physician's certificate of death. Of course none was forthcoming. Fullmer now began to recall for the first time past incidents that he said should have aroused his suspicions about Margaret's honesty, but which he confessed he either overlooked or managed to rationalize away.

A long letter from a former Rowenite and good friend of sister Rowen, Mrs. Grace McCausland, was printed. She recalled an incident involving her brother, a Campbellite preacher in the east, who refused to accept Saturday as the true Sabbath. Mrs. Rowen had a vision of this brother writing a letter in which he apologized for all the harsh things he had been saying about Adventist doctrine. Sure enough, a few days later Margaret showed Grace a penciled note from her brother containing humble apologies. However, the handwriting was not like the brother's, and the letter was not on his customary stationery. When Grace later visited her brother, she found him in no way apologetic. He vigorously denied sending the note. Grace added that she now realized that, when Margaret made a prediction that failed, she did her best to fabricate evidence of its fulfillment. "Cherish a spirit of pity for her," Mrs. McCausland adds, "instead of a spirit of condemnation," and pray for her deliverance from the Evil One.

The notion so long suggested by Fullmer, that February 6 was the date Jesus started on his journey to Earth, cannot be reconciled with Margaret's visions. Grace remembered that, on January 16, 1925, Mrs. Rowen claimed that Jesus himself said to her: "The 144,000, if necessary, will be taken from their abiding places and transported to the gathering place; and then immediately taken up in the clouds of heaven by the angels on February 6, 1925. These are my words and I your Redeemer." Clearly this vision left no doubt that February 6 was to be the date of the Lord's arrival, not the day of his departure.

Dr. Fullmer closed what I believe to be his final issue with an "Open Letter to Mrs. Rowen." In it he said he did not know of her whereabouts, although he was aware that she was traveling with at least one companion, Dr. Jacob Franklin Balzer, an Adventist physician who had become an ardent Rowenite.[2] Fullmer reminded Margaret of a time when she introduced him to a woman she claimed was her true mother. When he later visited this woman she vigorously denied it. Fullmer addressed Margaret:

> Regarding her acquaintance with you, she asserts this to have been of seventeen years duration: and that she met you when your family first came to Los Angeles and you were employed in a hatchery conducted by your sister, a very intimate friend of this woman. That same evening, at this woman's request, Mrs. Fullmer and I accompanied

her to your own sister's home, and received confirmation of the truthfulness of her story. How do you account for THIS?

Fullmer reminded Margaret that on three different occasions she gave three different dates on which she said she would be raptured. He added, with newfound sarcasm, that she must have been held down by the force of gravity. He informed her that he now had positive evidence that for years she had been stealing thousands of dollars in tithe money given by trusting members of the Reform Church. He recalled five expensive cars she purchased with those purloined funds. Margaret initiated two lawsuits against him and his wife, presumably for slandering her. Fullmer said he eagerly looked forward to her testimony under oath. His Open Letter ends with a plea to Margaret to "disconnect" from the satanic sources of her visions.

As we learned in the previous chapter, Margaret became so angered by Fullmer having turned against her that in February 1927 she arranged for Dr. Balzer and Miss Mary A. Wade, Balzer's nurse and friend, to try to kill Fullmer! He recovered from Balzer's blow to his head with a lead pipe. Mrs. Rowen, Balzer, and Wade were tried and sentenced to prison for attempted murder. Dr. Fullmer died of a heart attack before the end of 1927. What happened to Mrs. Rowen after her release from San Quentin, or what happened to her husband, George Rowen, and their three children is not known.

Rank-and-file Adventists today have never heard of Mrs. Rowen. Church leaders and historians who recall the Rowenite breakaway church see it as a comic example of the folly of setting a date for the Second Coming. They remember that their church had its origin in an earlier instance of just such a folly.

William Miller, whose fervid preaching about the Second Coming lead to the Adventist movement, was certain that the Lord would return sometime in 1843. The passing of this year was the first great disappointment among his followers. They quickly revised his arithmetic to settle on October 22, 1844, as the momentous date. Miller had been reluctant to name a month and day, but he finally approved of October 22 as the most likely. So positive were the Millerites that Jesus would return on that day that many paid off their debts, sold homes and businesses, and farmers left crops unattended.

No one has described their disappointment more vividly than Adventist historian Leroy Edwin Froom in chapter 38 of the fourth and final volume of his monumental *Prophetic Faith of Our Fathers*. Here is the chapter's last paragraph:

> From one home, as the day was ending and the Savior had not come, the sun was seen sinking over the western hills. Its last rays lighted up a cloud near the horizon, and it shone like burnished silver and gold. It was a glorious scene, and the father rose expectantly from his chair, thinking it might be the Savior coming. But it was only a "sun-kissed cloud," and the family resumed its waiting. Thus the day wore slowly on to its weary close, though far into the night the faithful kept vigil. But from those exalted heights they were soon dashed to the depths of despair. Their Lord came not, and the day of sweet expectation had become the day of bitter disappointment.

For a brief period James and Ellen White believed that October 22, 1845, would be the longed-for day. Hiram Edson, a Methodist turned Millerite, proclaimed 1850 as the year of the Advent. In 1850 Joseph Bates, one of the founders of Seventh-day Adventism—it was he who persuaded the Whites that God intended the remnant church to worship on Saturday—published a pamphlet that set off a new rash of date-setting. He argued that the Second Coming would occur in October 1851 because it was exactly seven years after the October 1844 date. However, before 1851 ended, this expectation was dashed by Mrs. White's June 21 vision condemning all date-setting. James

White wrote: "To embrace and proclaim a time that would pass by would have a withering influence upon the faith of those who would embrace and teach it."

Adventists now take pride in saying that their church no longer names a day, year, decade, or even a longer time period during which the Lord will return. They preach only that the end of history will be soon.

But how soon is soon? Few Adventists today are aware that Mrs. White, in her most famous work, *The Great Controversy Between Christ and Satan* (chapter 18, on William Miller), defends Miller's preaching that the great meteoric shower of November 13, 1833, fulfilled Jesus' prediction that the "falling of stars" would be a sign of his soon return. (See Matthew 24:29, Mark 13:25, and Revelation 6:13.)

After citing the falling of stars and the darkening of the sun and moon (which Miller and Mrs. White took to have been fulfilled by the famous Dark Day of May 19, 1780), Jesus added: "Verily I say unto you, This generation shall not pass, till all these things be fulfilled."

Exactly what did Jesus mean by "generation"? Liberal Christian scholars take it to mean the generation of those listening to him, and freely admit that Jesus either never spoke those words, or if he did he was mistaken. Adventists and other fundamentalists wiggle out of such heresies by assuming that by "generation" Jesus meant the generation of those who would witness the heavenly signs at some time in the future.

For decades Seventh-day Adventists taught that Jesus would return within the lifetime of at least some persons who saw the 1833 meteoric shower. After a century passed beyond 1833 with no Second Coming, the church began to expunge from their literature all references to the Dark Day and to the shower. Now that more than 160 years have gone by since 1833, the word *generation* must either be redefined by Adventists as an indefinite period of time, or the signs that Miller and Ellen White mistakenly took to be fulfillments of Matthew 24:29 are events that have not yet occurred. In any case, the Adventist church has finally learned that setting even a vague approximate time for the Second Coming must be scrupulously avoided.

In the third and final article of this series, I will tell the hilarious story of how a band of loyal Rowenites in New York handled the prediction that Jesus would return on February 6, 1925, the rapture of the 144,000 saints, and the horrendous earthquakes and plagues that would bring Earth's history to a close.

NOTES

1. Throughout the centuries there has been enormous controversy among conservative Christian theologians and leaders over the identity of the 144,000 saints. Seventh-day Adventist leaders have also differed sharply. For a three-page summary of their conflicting opinions see the entry on 144,000 in *The Seventh-Day Adventist Encyclopedia* (1966).

2. "Why the Use of Flesh for Food May Become Sin," a lengthy article by Dr. Balzer, attacking meat eating, appeared in *The Reform Advocate* (November 1923, pages 8–12). Fullmer later published it as a pamphlet.

THE COMIC PRATFALLS
OF ROBERT REIDT

> From time to time, as we all know, a sect appears in
> our midst announcing that the world will very soon
> come to an end. Generally, by some slight confusion
> or miscalculation, it is the sect that comes to an end.
> —G. K. Chesterton, in *The Illustrated London News*
>   (September 24, 1927)

In the first of three articles on Margaret Rowen I gave a
history of the Reform Seventh-day Adventist Church, a
breakaway movement started by Mrs. Margaret Rowen, a
Los Angeles housewife who claimed to be the God-chosen
successor of Ellen White. The history ended with Mrs.
Rowen and two accomplices going to prison for trying to
murder Dr. Bert Fullmer, her chief disciple.

My second article traced Dr. Fullmer's disenchantment
through the pages of his periodical *The Reform Advocate*.

This essay first appeared in *Free Inquiry* (spring 1997).

This, the third and final article of the series, is the comic account of Robert Reidt, a Long Island, New York, Rowenite and his little band of disciples.

No one knows how many Adventist followers Margaret Rowen managed to ensnare in the United States alone. Some historians estimate the number as about a thousand, others twice that many. Whatever the case, her influence stretched from California to New York, where it found its most vociferous disciple in the person of Robert Reidt.

The *New York Times* (February 5, 1925) described Reidt as a thirty-three-year-old German-born "pale-faced fat little man" with a "buxom" thirty-year old German wife, and four "pallid, frightened-looking little children." Calling himself the "Apostle of Doom," Reidt said he had been predicting the imminent Second Coming of Jesus and the end of the world for fourteen years. A former Seventh-day Adventist, he accepted Mrs. Rowen's date of February 6, 1925, for Jesus' arrival, not only because God had revealed this to her in a vision, but also because he himself had had similar visions.

As the hour of midnight approached, Reidt said, there would be a sign in the heavens visible only to the faithful— a little black cloud "no larger than a man's hand." It would be Jesus and a band of angels on their way here from heaven, a region just beyond the constellation of Orion. The cloud would descend on the Earth and transport 144,000 of the saved to a hilltop in the woods near San Diego.

During the next seven days, Reidt preached, the Earth would be destroyed by fires, diseases, pestilences, and hailstones. The unsaved would perish as mountains topple and bury them. Stars would fall from heaven as predicted in Matthew 24:29. The other sign predicted in the same verse, the darkening of the sun and moon, was taken to have been fulfilled by a total eclipse of the sun that took place on January 24. The eclipse had been visible in the eastern states just ten days before Reidt's Doomsday.

After the seven days of destruction, according to Reidt, God's powerful "searchlight" would shine on San Diego, and the 144,000 "brides of the Lamb" would be beamed up to start their journey to heaven. The trip would take exactly seven days. On the Sabbath (Saturday) Jesus and his band of angels and the redeemed would rest on the surface of Jupiter. Reidt was too ignorant of astronomy to know that Jupiter has no surface. How much of all this was taken from Mrs. Rowen's writings and how much added by Reidt I do not know.

Reidt and his family lived in an unpainted, ramshackle shack in East Patchogue, Long Island, New York. When a *Times* reporter interviewed him on the day before doom, twelve of his followers were praying softly in a dim back room. The *Times* described them as shabbily dressed, but with shining eyes. For the last few days they had been fasting and praying.

Reidt's wife and four children had no doubts about the approaching end of the world. The children were two boys, Robert, twelve, and Walter, nine, and two blue-eyed girls, Esther, six, and Ernie, ten. Taunts by schoolmates had made them miserable.

Reidt's top disciple, Willard G. Downs, of Yaphank, New York, was described as a "grizzled" fifty-seven-year-old former farmer and carpenter, tall, angular, with long hair and an unkempt gray-brown beard. His trousers were baggy. Ten years earlier he had converted from Methodism to Adventism. When he quoted Scripture his eyes flashed like lightning and his voice rolled like thunder.

Another Rowenite follower was Miss Katharine B. Kennedy, of Valley Stream, Long Island. She was described as a middle-aged spinster, who was a Baptist until seven years before, when she became a Seventh-day Adventist and a Rowenite. Other followers in Reidt's back room included Arthur Rupp, twenty-three, a recent Reidt convert, and a family of four blacks, also from Valley Stream.

Reidt claimed twenty-five disciples in Long Island. Most of them, including Reidt, were said to have sold their possessions so they could pay all their debts before being taken to Paradise. An earlier story in the *Times* (January 24) reported similar sales of homes and property by Rowenites in Washington state, Baltimore, Boston, Los Angeles, and other cities. In Greece, New York, a town near Rochester, farmer Elzear T. Smith killed his pigs, sold his cow and furniture, and "wound up all his earthly affairs" to await the end of the world.

On February 3, the *Times* printed a letter from Carlyle B. Haynes, a prominent Seventh-day Adventist, in which he referred to the Rowenites as a small fanatical group totally rejected by the Adventist church. The church, wrote Haynes, did not set a date for the second advent of Jesus, but only preached that it is "immanent." On February 6 the *Times* printed a letter pointing out that, contrary to what Haynes had written, Adventists had set dates in the past, and that the church's views were responsible for Rowenism.

On February 6, the day of doom, the *Times* reported that hundreds of persons had motored to Reidt's tumble-down shack in East Patchogue to see how he and his disciples would react if nothing happened. Brother Downs is said to have sold his last possession, a bicycle.

Reidt emerged from his shack to announce that the world would not end at midnight. At midnight he and followers would see in the east the little black cloud that indicated Jesus and the angels were on their way. It would take seven days for them to reach the Earth. A week of terrible destruction would then occur, and the 144,000 saints would take seven more days to get from San Diego to heaven. There would be many stops along the way to pick up saints from other planets.

Reidt had recently shaved and was wearing neatly pressed trousers. Brother Downs was "resplendent" with a new blue necktie. Mrs. Reidt had on a brown silk dress.

Downs announced that Jesus would be escorted by angels, led by Gabriel. They would march on Earth with feet of fire, and the Earth would be filled with music never heard before. To emphasize the role to be played by music,

Reidt produced an old zither on which he strummed, saying he would be playing it soon for the angels. His music, the *Times* said, "was not so good."

On February 7, the *Times* reported the events of the previous night. Reidt not only had failed to see the little black cloud, he could not even locate Orion. There was only a bright, gibbous moon rising in the east, waning from the total eclipse it had produced eleven days earlier. A subdued Reidt faced a battery of more than a hundred persons, standing in the mud and snow surrounding his shack. Numerous reporters and motion picture camera men were on hand. A tired Reidt retired at midnight, saying he would address a meeting at 10:00 the following morning. Brother Downs was in a back room arguing about Scripture with a young Italian Roman Catholic. Some prankster boys in the neighborhood caused a flurry of excitement when they burned a cross in the woods nearby.

The *Times* painted a dismal picture of the scene. "The shack is two miles from any house. It stood alone, melancholy and ghostly, in fields of dwarf oak and winter-blighted shrubs. Every window was curtained and the house might have been tenantless except for the fragments of prayers that came to those waiting in the moonlight outside."

"Oh, Jesus, we are ready," were the words of Reidt himself.

The next morning about 150 people showed up at the shack, including a policeman assigned to keep order. Reidt's twelve disciples were nowhere to be seen. He said they were home awaiting the arrival of Jesus in seven days. Reidt said he had sold his shack, his furniture, and his Ford car. The car had foot-high letters painted on its sides announcing the world's end in February. It was rumored that the sales were contingent on the end of the world occurring.

On February 7 the *Times* interviewed Mrs. Rowen's mother, brother, and other relatives living in or near Philadelphia. They had moved there a few years earlier from Los Angeles. The mother, Mrs. Mathilde Wright, said that, when she last visited her daughter in Hollywood in 1923, she was appalled by her daughter's visions and prophecies, and thought she should be spanked. Margaret's three children had been completely won over by their mother's views. Mrs. Wright felt sorry for them. Whenever they saw a dark cloud they would run to her and say, "Granny, does that mean the world is coming to an end soon?"

Mrs. Wright called herself a good Methodist. She described her daughter as "not four feet tall and almost as wide." Nothing was said about Mr. Rowen. He remains a curious blank in our story. I do not even know how he made a living.

Six other brief news items about the Rowenites appeared in the February 7 *Times*. Another Adventist spokesman denounced Rowenism. Miss Katherine Kennedy, of Valley Stream, was quoted as saying that, if nothing happens in the next week or two, she would lose her faith in Sister Margaret. The six Valley Stream area Rowenites included one Herman Steubbe, a fifty-year-old Methodist living on Washington Avenue in nearby Roosevelt.

Karl Frederick Danzeisen, a forty-nine-year-old farmer in Temperance, Michigan, was reported to have shot his wife, then killed himself with another bullet. The recovered wife said he was "terror stricken" over the world ending on February 6.

In Lincoln, Nebraska, twelve Rowenites were said to be awaiting the black cloud and the searchlight. They had sold all their worldly goods. Mrs. Rowen, at 112 Gower Street, Hollywood, was sticking by her prophecy.

On February 8, the *New York Times* revealed that Brother Downs had had a dream in which an angel, perhaps Gabriel, told him that believers in North and South Dakota had seen the little black cloud. Why had he and Reidt not seen it? Brother Downs had the answer. Satan had used the magnesium flares of motion picture cameras to blot out the eastern sky.

On the previous day, Downs had become so angry with the camera men that he threw water over one of them, but now he and Reidt were agreeable to more movie making.

For the cameras, Downs hung around his neck a sandwich board that said: "Prepare to meet thy God. Seventh-day Sabbath is the Seal of God. Sunday is the Mark of the Beast. Amos 3:7." These statements were all good Seventh-day Adventist doctrine.

Reidt and his family vanished on February 10 from East Patchogue. No one saw them leave. He had been hissed the night before when he spoke on the stage of a local theater. The orchestra had added to the hilarity by playing the song "California, Here I Come."

On February 12, the *Times* learned that Reidt and Brother Downs had popped up in Newark, New Jersey. The apostle of doom paraded down a business street wearing a sign announcing the coming end of the world. He now predicted this would occur before the end of 1925, probably in September. He said he left East Patchogue because of threatening letters, giving up a good job in the painting and decorating business. A photograph on the front page of Newark's *Star Eagle* (February 11) showed the little prophet with a leather Bible in his left hand, his right arm upraised, and his "blue eyes shining." He had visited several newspaper offices in Newark, but refused to say how long he intended to stay in town.

The Newark *Evening News* of the same day reported that the apostle of doom was penniless. He and his wife and four children were hiding out in two furnished rooms, still fearful of death threats. "His pockets stuffed with tracts and booklets of the Reform Seventh-day Adventist doctrine, a battered, cheap Bible next to his heart, Reidt is a pitiful figure. His clothes are shabby, but he still expects to exchange them some time soon for the pure white robes of an angel. Meanwhile, friends who remembered him when he taught a German Adventist Sunday School class in Newark prior to 1920 are aiding the destitute family."

The unsinkable Reidt did not surface again until January 2, 1926, when the *New York Times* disclosed that he had sent a letter to New York City's Mayor Jimmy Walker.

"APOSTLE OF DOOM" SEEKS SANCTUARY HERE

PROPHET REIDT NOW SAYS WORLD WILL COME TO AN END IN SEPTEMBER

Robert Reidt, the "Apostle of Doom," as he appeared on the front page of the Newark, New Jersey, *Star-Eagle*, on February 11, 1925.

It requested permission to broadcast a new date for a divinely scheduled calamity. However, instead of the entire world being destroyed, Reidt now predicted that on February 6 New York City would be destroyed by fire from heaven. Evidently Reidt had decided that Mrs. Rowen was right about the month and day, but wrong about the year and the extent of the doom. Nothing was said about the Second Coming of Jesus.

Reidt's letter to Mayor Walker included these stirring words:

> I, as far as personal comfort and safety are concerned, would rather be pursuing my trade as a painter and deco rator than to hear the mocking of the unbelieving world and suffer the reproach that is my daily lot. But I love my fellow man, no matter of what creed and color. I know that I will be a lost man for all eternity if I will not do all in my power to warn the people of the destruction that is so sure to come.
>
> We do not want the blood of a single person on our con- science. Convinced in our very souls that it is a message from God, we have promised our Creator to do our level best in warning the people of New York. We would be glad for any assistance that your esteemed office could render. We want to reach the people some way or other. We are not after money or honor. These things we are and always have been willing to obtain by honest toil. We beg you to make it pos- sible for us to use the city's broadcasting station or any public place or park for the purpose of reaching the masses.

I leave it to the reader to decide if these are the words of an honest but deluded man or the remarks of a charlatan.

Two days later the *Times* reported that Rowenite Collin N. McCloud, a member of the Reform Seventh-day Adven- tist church and one of Reidt's apostles, was arrested in Roosevelt, Long Island. He had been driving Reidt's old bat- tered car, the so-called Chariot of Doom, without 1926 license plates or a driver's license. Reidt was said to be living in Baldwin, Long Island, apparently alone. McCloud's address was 6 Charles Street, in Roosevelt.

McCloud had been arrested at 2:30 A.M., when two Roo- sevelt patrolmen heard the loud noise of the "chariot" rat- tling through the town's quiet streets. The apostle was accompanied by two teenage girls who said they were just "joy riding." McCloud spent the night in jail. The car was found to belong to a man named Christian who lived in Brooklyn on 47th Street. Reidt showed up at the police station to obtain his car, which the police had impounded. He claimed he had bought the car two months earlier, but was unable to produce a bill of sale. Before he left to get the bill, a police lieutenant said, "You don't need the car any more because you said you were going up to heaven in a chariot of fire on February 6, and that's only a few weeks off."

According to the *Times* of January 27, Reidt intended to broadcast his warning over a radio station in Bay Shore, Long Island. Two days later the newspaper revealed that he and two disciples had strode into City Hall to warn Mayor Walker that on February 6 the city would first be leveled by "great earthquake," followed by seven days of terrible conflagration. The apostle of doom was now said to reside in Baltimore, Maryland.

Mayor Walker was too busy to see Reidt. While the prophet was standing in a corridor, hoping to see the mayor, Democratic Alderman Charles A. McManus asked Reidt to postpone the date of the Big Apple's destruction because February 6 was his birthday. "You'll have a chance to celebrate during the fire," Reidt replied.

Of course nothing happened on February 6. The rotund Reidt, after discovering errors in his calculations, revised them for the third time. On February 11 he was back in Manhattan to speak over radio station WRNY, at the Hotel Roosevelt, warning that the city would be destroyed shortly after midnight by a huge ball of fire that would fall from heaven. Everyone would be killed within a twenty-five-mile radius. Reidt said his new prophecy was based on the Old Testament's Book of Ezekiel. "I wouldn't give five cents for the Woolworth building," he said. By now he had become a figure of fun whose predictions frightened nobody. Three strikes and you're out.

A *Times* headline of February 14 was "Reidt Wrong Again on Doom Forecast." He is said now to be back in

East Patchogue. There is no mention of his family. Was he living alone? Nothing more about him appeared in the *Times*. It would be interesting to know what finally happened to the crazy little Rowenite and to his long suffering wife and young children.

JOHN MARTIN'S BOOK

A Forgotten Children's Magazine

Afew thousand older Americans from middle and upper income families have fond memories of a monthly magazine to which their parents subscribed when they were very young. It was called *John Martin's Book*. You'll not find it mentioned in any history of children's literature, although five pages are devoted to it in *Children's Periodicals in the United States*, a valuable reference edited by R. Gordon Kelly (Greenwood 1984). Even oldsters who read it sixty years ago or had it read to them can tell you nothing about John Martin or the history of his remarkable periodical, yet in its time it was the most entertaining magazine published in this country for boys and girls aged five to eight. In many ways it was a pioneering publication.

There have been more than four hundred periodicals in the United States for young people of varying ages. The

This essay first appeared in *Children's Literature*, vol. 18 (1990).

first, *Children's Magazine*, was born in Hartford, Connecticut, in 1789 and died three issues later. By far the most influential was *St. Nicholas*, founded in 1873 by Mary Mapes Dodge, who is best known for her novel *Hans Brinker or the Silver Skates*.[1] The only child's magazine of comparable literary quality in the United States today is *Cricket*. A glance through any issue of St. Nicholas from its golden years shows that it was intended for readers older than eight. After 1900 it even began to run articles for parents, including interviews with William Gladstone, Thomas Hardy, William Dean Howells, Henry Ward Beecher, and other notables who had nothing to do with children's literature. After Miss Dodge's death in 1905, *St. Nicholas* fell to a succession of owners and editors before it expired in 1943 as a picture magazine sold in five-and-dime stores. During the first two decades of this century, older children could still obtain the famous *Youth's Companion*. *American Boy* and *Boy's Life* were also available for older boys, but there simply was no magazine of quality for very small children. Even the monthly *Child Life* appealed to youngsters over the age of eight.

It was into this vacuum that *John Martin's Book* (hereafter *JMB*) entered in 1913. The key to this magazine's success was the unfeigned delight taken by its publisher and editor, and by his associate George Carlson, artist and puzzle-maker, in the child's intellect and imagination. They deemed the ability to "play" with one's mind worthy of adult respect. In 1921 (I was seven and a subscriber) I wrote to Martin for his autograph. He responded with typical zest by saying he would rather send it to me than to a king. More than thirty years later, when I was a contributing editor for eight years to *Humpty Dumpty's Magazine*, my main task was inventing what I called "gimmick pages"—activity features that involved cutting, folding, writing on, and otherwise damaging pages. For each issue I also wrote a short story, designed to be read aloud by an adult, about the adventures of Humpty Dumpty Junior.

Junior, a small egg, was the son of the magazine's supposed editor, Humpty himself. I also wrote for each issue a poem of moral advice spoken by Humpty Senior to his son.[2] Memories of *JMB* were my inspiration for the activity features I created. I took up, so to speak, where Carlson left off. years later, when I began collecting issues of *JMB* (many of which I had not seen as a child), I was amazed by how many of my ideas Carlson had anticipated. I became increasingly curious about both Martin and Carlson, a pair who had so early addressed themselves to the imaginative needs of very young children and had brought some of the best freelance writers and artists of the day into the venture with them.

John Martin, I discovered to my surprise, was a pseudonym. His real name was Morgan von Roorbach Shepard. What little is known about his life comes entirely from two sources: an interview with Allan Harding in *American Maga-*

John Martin, as he appeared in a Player's Club production.

zine (August 1925) and three pages by Martin himself in the July 1923 issue of *JMB*. Although Morgan was born in Brooklyn in 1865, his first nine years were spent in poverty on a plantation in Maryland. His mother had married at age sixteen. "She was my playmate, teacher, and mother—all in one," Martin told his interviewer.

> She and I lived in a world of our own; a world where fact and fancy went hand in hand. For instance, in the yard there was a bird house, occupied by a colony of martins. To me they were as real as a human family. . . . My mother talked of them by name: John and Joan, Robin, Alice, and a dozen or so more. John was the leader bird, and their house was John Martin's House. When the birds would return to their home from mysterious flights they brought back tales of adventure which Morgan's mother would tell her son. That was the way my mother taught me geography. Sometimes they were stories of animals and birds and fishes, and so I learned natural history. Or tales of heroes and people who had lived a long time ago, and that was my introduction to human history. And best of all, these magic martins were intimately acquainted with fairies, and not at all averse to letting me know their secrets.

Morgan had nothing to say about his father and I have been unable to learn anything about him. The death of his "beautiful girl mother" (as Morgan called her in 1923) when he was nine was a crushing blow; in a sense he never got over it, or the abrupt changes it brought about. "For several miserable years I was in a boarding school where I was man-handled, bullied, and misguided almost beyond endurance. I was frail in body and sick at heart." Goaded by desperation, he would fight back with "blind fury," tossing books and bottles, overturning chairs and tables. Off he was packed to another boarding school where he was better treated, though his loneliness and unhappiness persisted. In 1881 the sixteen-year-old was "dumped out into the world." For a long time he made what he calls "a queer hash of it."

Morgan took off for South America but got no farther than Central America, where he found himself involved in a revolution. He took the side of the "outs" and for a while was active in their cause, though he does not disclose either the cause or the country. Details about this phase of his life are unknown; indeed, it is not even known whether we can trust everything Morgan told his interviewer. From Central America, Morgan went to California, where he made soap-box speeches "always against somebody or something that was 'in' at the time." Years later, he said, the child readers of his magazine cured him of "that particular sickness, just as they have cured me of bodily ills."

In California, Morgan moved restlessly from job to job. "I worked in mines, punched cattle, herded sheep, dragged a chain for a surveying party, oiled engines, picked grapes, and bucked wheat sacks in the wake of a harvesting machine. All this . . . hardened me outside, and maybe softened me inside." He was fired as a streetcar conductor for giving free rides to children. After a short stint as a newspaper reporter, he worked as a bank clerk for more than thirteen "interminable years." Morgan likened himself in this hated job to "a man pulling an oar in a slave galley." A nervous collapse ended this phase of his life.

In partnership with a book dealer Morgan then started a small publishing house in San Francisco but gave it up in 1904. After a period in Europe, draining the money he had saved, he returned to San Francisco where he opened a business designing greeting cards. The Crocker Building in which he had his office, was demolished by the great 1906 earthquake, and with it the business. Morgan thereupon moved to New York City.

While recovering from an operation on his leg (it had been severely injured when he tried to retrieve possessions from his office after the quake) Morgan began to write and sell poems to children's magazines. Ill-educated and unlucky, torn between rebelliousness and the need to earn a living, he found continuing pleasure in writing long letters to children. The letters were signed John Martin, and henceforth we will call him by that name. Like Lewis Carroll, who also wrote for children under a pseudonym, Martin had a mild talent for drawing; unlike Carroll, his love of children extended to both sexes, not just to little girls. His letters would swarm with funny little pictures that illustrated whimsical stories. When he was with children he liked to ask them to make squiggly lines on paper while he jiggled their elbow, then he would add more lines to create what he called a "quiz-wiz" animal. In the twenties, *JMB* would run quiz-wiz competitions, giving prizes to children who sent the funniest drawing of an animal based on a published wiggly line.

In 1908 Martin asked himself: why not turn my letters into a small monthly periodical? With the addresses of a few hundred mothers, many of them supplied by his sister in Orange, New Jersey (I do not know if he had other siblings), he began to solicit subscriptions. The first printed letter, handwritten and illustrated by Martin, went to four subscribers. Soon he was mailing out two thousand.

Printed on tinted paper that varied in color from letter to letter the contents were intensely personal, loving, sentimental, informal, chatty, and pious. Each letter opened with "My Dear . . ." followed by a space in which Martin would add the child's name. "How do you do?" the first letter began. "I hope you are very well, and as happy as can be. My name is John Martin. I love *little* Boys and Girls and I *play* with them, too. I have had quite a good many birthdays but I never grow very old—my heart stayed young. I can run fast and if I tumble down I get up and laugh some."

One finds in these letters almost all the ingredients that would later go into *JMB*. There are sentimental poems and tales about the beauty of nature, fairies, pets, mermaids, and knights, all with a strong emphasis on love and religious faith. It must be admitted that Martin's verse was doggerel on a level below that of Edgar Guest, but a child could understand every word. God is often invoked, as in Martin's

John Martin

most frequently anthologized poem, "God's Dark," though seldom with reference to any particular faith.[3] For moral instruction the letters introduced a family called the Chubbies, modeled on Gelett Burgess's popular books about a family called the Goops. The Goops were ill-mannered youngsters whose beastly habits were described in verse by Burgess, of purple-cow fame. By contrast, Martin's Chubbies were good, well-behaved, plump little children who became a staple feature of *JMB*. The following jingle by Martin accompanying a picture in the August 1919 issue of moon-faced Chubbies going off to school, is typical:

> The Chubbies love the life they live
> And all the goodness in it,
> And they are always punctual,
> Exactly on the minute.
> They never waste my time or yours
> And make us wait or worry.
> But still they don't get out of breath
> With needless rush and hurry.

Martin's letters also anticipated the activity features that would play such a major role in *JMB*: secret codes, riddles, rebuses, puzzles, and pages to be folded or cut. A 1910 letter introduced what mathematicians call a dissection puzzle so clever that it would be used scores of times in later years as an advertising premium. There are four polygonal shapes to be cut out and formed into a large T. "If you can do this puzzle in 10 minutes let me know," Martin wrote. He urged his readers to trace the shapes on transparent paper, paste them on white cardboard, and carefully cut them out. "Of course you can cut the pieces out of my letter, but that would make it all holey and raggerty and words on the other side of your puzzle page would be all gone. . . ." This emphasis on the pleasure a child gets from mental effort was more than guesswork; Martin always kept in close touch with his readers. Throughout his years as editor of *JMB* he spent "working vacations," as he called them, on the beaches of

Nantucket, where he sat under a large blue umbrella with the initials "J. M." on it. He liked to hide 500 Chinese coins in the sand, offering a prize to the child who found the most. He would give away 500 free tickets for ice-cream cones. "It isn't good sporting to ask for these tickets," he said in his 1925 interview. "Children are natural grafters but they are better natural sports if you'll give them a chance. If a child *asks* for one, he doesn't get it. Five hundred goodly doings that are worth something get as many cone tickets out of me."

After four years of success, Martin changed the name of his periodical from *John Martin's Letters* to *John Martin's Book*. The first issue appeared late in 1912, published by John Martin's House, 5 West 39th Street, Manhattan. (A British edition was distributed by G. Bell and Sons.) Pre-1920 issues are now extremely rare. The earliest in my collection, October 1913, has 96 pages on heavy stock, without pagination. Beneath the title are the words "A magazine for little children." The price is 25 cents.[4] The cover picture of Eeny, Meeny, Miny, and Mo is by Martin; beneath is the couplet:

> All Life Is Full of Fun—Hurray!
> Who Cares For Care? Let's Laugh and Play.

Those lines set the tone for all the later issues. No editor ever cared less about winning the praise of librarians. "Children want someone who will play with them," Martin said in his interview. "They want to play in their minds and their imaginations, just as my own wonderful mother played with me. . . . Most grown people have sat on their imaginations so long and so hard that they have crushed the life out of them. A child's beautiful world of Make-Believe is a place they don't even try to enter. Many children, knowing that average adults do not understand, live a whole secret existence in the world of Make-Believe. They are alone and unguided there. Yet it is the one realm where they can be most receptive and most responsive." Martin saw his readers not as romantic children trailing clouds of glory but as children whose secret world was a region that should be expanded by loving adults who could share it and direct it as his own mother had done. The stories in the October 1913 issue range from realism to pure fantasy. Martin contributes an article that purports to be a letter from his dog Rubber, explaining how to tell what a dog is feeling by the postures and movements of its tail. This is followed by Martin's poem against cruelty to pets, and a decalogue of commandments on how to treat your dog. The issue reprints the T puzzle Martin had introduced three years earlier in one of his letters. Cut-outs such this, which would become increasingly abundant in later years, show how little concern Martin had for library sales. No library wants a magazine that is likely to be mutilated, but Martin did not mind in the least. Nor did he seek income from advertising products he disliked. All the ads were written by Martin, or under his supervision, and in a way that made them seem like pages that were not advertising. He refused all ads for chewing gum, a habit he considered harmless but vulgar.

From the outset Martin had the able editorial assistance of Helene Jane Waldo, who remained with him until the magazine's final issue in 1933. *JMB* certainly did not compare with *St. Nicholas* in its number of famous writers, but there were a few: Thornton Burgess, author of many books of animal tales; Conrad Richter, a New Mexican novelist; and Grace Adele Pierce, whose book *The Prairie Queen* reprinted stories she wrote for *JMB*. We should realize, however, that famous authors seldom write for five-year-olds; contributors of fiction to *JMB*, though not distinguished, were probably as good as any available at the time for the magazine's age level.

The artwork was on a higher level. Among the leading graphic artists who drew for *JMB* were the Gruelle brothers. Johnny Gruelle was the author and illustrator of the popular Raggedy Ann books, and his brother Justin contributed even more frequently to *JMB*, not only art but also stories and verse. Jack Yeats, brother of William Butler Yeats, wrote and illustrated pirate tales. William Wallace Denslow, best known for his color plates in the first edition of L. Frank Baum's *Wonderful Wizard of Oz,* drew pictures for poems that appeared in the July and September 1915 and September 1916 issues.

Another frequent contributor to *JMB*, both as a writer and artist, was Frank Verbeck, who illustrated Baum's *The Magical Monarch of Mo*. In the April 1927 issue Wanda Gag—she became famous the following year for her best-seller *Millions of Cats*—published a clever story, "Bunny's Funny Easter Eggs," in which italicized words were used for solving a crossword puzzle.

The most important artist associated with *JMB* was George Carlson, a prolific illustrator who had been trained at the New York Academy of Design. A list of his books would run to almost a hundred titles. Dozens of them were novelty paperbacks, many published by Platt and Munk. They included coloring books, paint with water books, connect-the-dots books, maze books, how-to-draw books, a series about Uncle Wiggily, and books of riddles, games, and crossword puzzles.

Carlson also drew for *St. Nicholas, Youth's Companion, Judge, Scribner's Magazine*, and other periodicals. He was responsible for many of the hidden-picture pages in *Child Life*, and for twelve years he was puzzle editor of the Girl Scouts' *American Girl*. He had a regular puzzle page in a comic book called *Famous Funnies* and contributed features to some forty issues of *Jingle-Jangle Comics*. The two published issues of his own *Puzzle-Fun Comics* (1946) are now extremely rare collector's items.[5]

Here is a partial list of the juvenile fiction he illustrated: Chandler Oakes's *Toby Town*; Gene Stone's *June and the Owl* and *Adventures of Jane*; Johanna Spyri's *Toni the Little Woodcarver* and *Tiss, a Little Alpine Waif*; John Martin's five *Read Out Loud Books*; J. L. Sherard's *Blueberry Bear* and *Blueberry Bear's New Home*; Mary

George Carlson. Note his jacket for *Gone with the Wind*, and his comic books on the wall.

Patterson's *Rip Van Winkle* and *The Legend of Sleepy Hollow* (retelling Washington Irving's stories); and Mark Twain's *Tom Sawyer*. In my opinion his most impressive work was the set of full-color plates tipped into Elizabeth Blanche Wade's novel *The Magic Stone*. Among the jackets he designed for adult novels was the jacket for the first edition of Margaret Mitchell's *Gone with the Wind*, still seen today on hardcover editions. Twenty-two years younger than Martin, Carlson was 75 when he died in 1962.

When *JMB* was first started, Carlson was merely a friend of Martin's and an occasional illustrator for the magazine. Soon he became the magazine's most frequent cover artist, drawing more than fifty covers. Carlson also created almost all the magazine's puzzles, activities, jokes, and riddles. He was responsible for an enormous variety of "gimmick" pages of a sort never before attempted in a child's magazine. There were pictures that turned into something else when you inverted the page (see figure 2). There were optical illusions, shaped poems, cut-outs that cast startling shadow pictures on the wall, stories with blanks in which children put their own adjectives. There were pictures that changed in funny ways when the page was folded and hundreds of cut-outs. Little doors would open, strips would slide back and forth through slots, disks pinned to a page would rotate to make amusing changes.

There were pictures with captions that could be read by holding the page to a mirror. There were instructions for folding origami animals and for making ink-blot pictures and simple cardboard or wooden toys. There were connect-the-dots rebuses, anagrams, ciphers, puns, crossword puzzles, science experiments. Walter Gibson, who later wrote books about magic and created radio's famous mystery series about the Shadow, supplied mazes that formed animals when you drew the correct path.

My own favorite as a child was Carlson's monthly page called "Peter Puzzlemaker" (see figure 3). Martin introduced the feature (October 1918) by saying that he sometimes had to tell Peter his puzzles were too hard. "We want our puzzles to be just hard enough to make you work over them, but easy enough to solve before you get fidgety and impatient." In addition to a simple puzzle, each picture contained a mistake that you tried to find before the next issue revealed it. The mistakes were amusing and clever: a keyhole upside down, tallow dripping upward on a candle, a rake's handle that went behind a fist instead of through

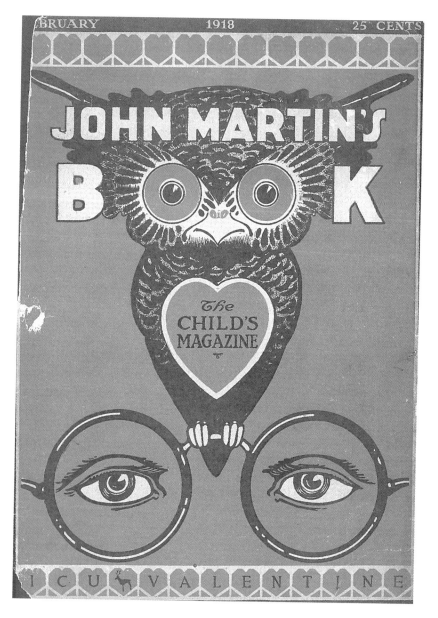

Figure 1. In this cover for the February 1918 issue, Carlson used the two Os in "Book" to form the owl's eyes. Observe also the word play at the bottom: "I see you dear Valentine." John Martin's covers were usually topical: Halloween for October or rain for April.

Figure 2. Carlson drew dozens of pictures for *John Martin's Book* that changed to something else when turned upside down. In this drawing, from the March 1920 issue, he has lettered "moo-cow" in such a way that it remains the same when inverted. It is an early example of a calligraphic technique refined by Scott Kim in his beautiful book *Inversions*.

Figure 3. This Peter Puzzlemaker page is from the May 1922 issue. The original caption told readers that "a charade is a sort of puzzle in verse." "Read this one carefully," it suggested, "and see what words Peter has in mind. This is a very old-fashioned charade, more than a hundred years old. Jack-in-the-box says there is a mistake in the picture, too." Answers were given in the following issue.

it, a star inside a crescent moon, an extra finger on a hand, smoke and a weather vane showing the wind blowing opposite ways, a cat without whiskers, a book with its title on the back cover, and so on.

A collection of these puzzle pages were issued by John Martin's House in 1922 as *Peter Puzzlemaker*. Readers were asked to cut out a large paper padlock from page 7 and paste it over the edges of the answer section to prevent peeking. No better collection of puzzles for young children was ever published. Why has it never been reprinted? The material remains fresh, and even Carlson's cartoon style has a refreshingly quaint period quality. It is unobjectionable to today's children who, like children in all ages, are indifferent to fashionable art trends. If a reprint sold well, it could be followed by a dozen other books of later Peter Puzzlemaker pages.

Throughout its history John Martin's House published many hardcover volumes, notably a series called *John Martin Big Books* that reprinted stories and articles from the magazine. Activity features were also reprinted in hardcovers, impossible to find today because most of them were designed to be destroyed. Titles include *The Fold-Up Book*, *The Jolly Book*, *Handy Hands Book*, *Something to Do Book*, *Some Fun to Make Book*. A volume called *The In-and-Out and Up-and-Down Book* had holes in the pages, tiny doors that opened, and a text that ran down one page and up the other.

When Martin was interviewed in 1925 he said his magazine's circulation was 40,000. "I don't make money out of it. Most of my alleged salary as editor is turned back at the end of the year to help cover the customary deficit. I am the richest and happiest man on earth, however, for my ledgers show a big profit in joy giving and getting."

In July 1928 the format of *JMB* expanded from $7\frac{1}{4}$ by 9 inches to 9 by 12, and pagination was adopted for the first time. In 1932 Martin moved his cluttered office to Concord, New Hampshire where his magazine expired the following

An array of *John Martin's* magazines.

year. A story in *Time* (August 1, 1932) entitled "Child-Man" had announced that magazine publisher George T. Delacorte Jr. planned to issue a new publication called *The Children's Magazine*. It would be for readers aged five to eight, sell for ten cents in the Kresge and Kress chainstores, and have John Martin as editor. I do not know if this magazine ever appeared, or if it did, whether any copies have survived. *Time* described Martin as "small, wiry, baldish" and a chain-smoker. "He speaks confusingly about himself as a dual personality: 'altruist, idealist, and hardboiled, almost unmoral.'"

John Martin died in 1947 at the Player's Club in New York City, of which he had long been a member. He was 82. An obituary in the *New York Times* disclosed that in 1900 he married Mary Elliott Putnam who died in 1942. In spite of his often cloying sentimentalism, his pious doggerel, and his frequent "writing down" to readers, Martin had great insight into a small child's interests and sense of fun; above all, he had the wisdom to give his friend Carlson free rein. He had no children, except of course the scores of thousands who enjoyed his magazine, among whom I unashamedly count myself.

NOTES

1. See Paul Rossa's article on *St. Nicholas*, "The Magazine That Taught Faulkner, Fitzgerald, and Millay how to Write," in *A American Heritage*, December 1985.
2. "Never Make Fun of a Turtle, My Son" (Simon and Schuster, 1969) was a collection of these poems, with Junior's name altered to Tom or Rose.
3. The poem can be found in *God's Dark, and Other Bedtime Verses and Stories* (1927). Martin wrote at least two other books of verse: *Aesop's Fables in Rhyme* (1924) and *A Jolly Big Alphabet* (1913).
4. The imaginative format of this issue—large type, lots of "air" on the pages, no page numbers, and one-color overlays (the color varying from issue to issue)—remained unchanged until

1928, when the magazine went to an even larger size. No format could have been less like that of *St. Nicholas* where the type became smaller color was diminished, and photographs started to appear.

    5. On Carlson's comic-book work, see Harlan Ellison's essay "Comic of the Absurd," in *All in Color for a Dime* (Ace Books, 1970), edited by Dick Lupoff and Don Thompson.

POSTSCRIPT

In 1992 Dale Seymour Publications reprinted *Peter Puzzlemaker*, a collection of George Carlson's puzzle pages that John Martin's House published in 1922. In 1994 I edited for Dale Seymour a new collection of Carlson's puzzle pages for a companion volume titled *Peter Puzzlemaker Returns!* I wrote an introduction about Carlson, and included some of his pictures that turn into a different picture when turned upside down, as well as pages of riddles that he illustrated. There are enough Peter Puzzlemaker pages in later issues of Martin's magazine to make a dozen other books.

    Twenty color pages from Carlson's *Jingle Jangle Comics* are reproduced in *The Smithsonian Book of Comic Book Comics* (1982), edited by Michael Barrier and Martin Gilliams.

WORD PLAY IN THE FANTASIES OF L. FRANK BAUM

A part from the famous Oz books, L. Frank Baum wrote many fantasies for children that did not have Oz as a setting. In all these books, he liked to make up names for persons, places, and things. These names often involved word play. Puns were the most common form, but Baum also indulged in anagrams, spoonerisms, reversals, and other forms of linguistic whimsies.

I have recently compiled an alphabetical list of all of Baum's invented proper names, and tried to identify the intended wordplay. In most cases, Baum's made-up names have no special significance; when word play is involved, it is most often obvious. Examples include such names as Betty Blithesome, Madame de Fayke (a psychic donkey who reads hoofs!), Tottenhots, Professor Nowitall,

---

This essay first appeared in *Word Ways* (May 1998) and was reprinted in *The Baum Bugle* (autumn 1998).

Nuphsed, Ann Soforth, Tom Ato, Grunter Swyne (a pig), Miss Trust, Duchess Bredenbutta, and hundred of others.

Here are some alphabetized instances, of less obvious word play:

Sir Austed Alfrin, a poet (England's poet laureate of the time was Sir Alfred Austin), *John Dough and the Cherub.*

Sir Pryse Bocks, "surprise box," *John Dough and the Cherub.*

Chopfyte, a composite man made of parts of the bodies of Nick Chopper (the Tin Woodman's name before he became tin) and Captain Fyter, *The Tin Woodman of Oz.*

Miss Cuttenclip, who cuts and clips paper figures that come to life, *The Emerald City of Oz.*

Dama, a forbidden fruit (a reversal of Adam), *Dorothy and the Wizard in Oz.*

Johnny Doit, a carpenter who can make anything in a jiffy, *The Road to Oz.*

Dot and Tot, the protagonists of *Dot and Tot in Merry-land* (Dot for Dorothy, Tot for Toto).

Joe Files, an army private (extensive files are necessary in the army, and soldiers march in files), *Tik-Tok of Oz.*

Glinda, the good witch of the Quadling region of Oz (an anagram of the last six letters in Quadling), *The Wonderful Wizard of Oz.*

Hartilaf, a giant with a hearty laugh, *The Magical Monarch of Mo.*

Herkus, persons of great strength like Hercules, *The Lost Princess of Oz.*

Inga, prince of the Island of Pingaree (Inga Consists of letters two through five in Pingaree), *Rinkitink in Oz.*

Jellia Jamb, Ozma's maid (jelly or jam?), *The Marvelous Land of Oz.*

General Jinjur, a woman full of "ginger," *The Marvelous Land of Oz.*

King Krewl, a cruel king, *The Scarecrow of Oz.*

Kwytoffle, a quite awful humbug wizard, *The Enchanted Isle of Yew.*

Princess Langwidere, a woman with a languid air, *Ozma of Oz*.

Li-Mon-Eag, a hybrid monster combining lion, monkey, and eagle, *The Magic of Oz*.

Noland, a mythical "no land," *Queen Zixi of Ix*.

Oz, it has been widely reported that Baum saw O-Z on a file cabinet, but this may be a myth. If O and Z are shifted backward one step in the alphabet you get NY, the abbreviation of the state where Baum was born and reared. If O and Z are shifted forward one step, with Z joined to A, you arrive at PA, the state where Baum's successor, Ruth Plumly Thompson lived, *The Wonderful Wizard of Oz*.

Princess Ozma, the appended "ma" could stand for mother or the first two letters of Maud, the name of Baum's wife, *The Marvelous Land of Oz*.

King Phearce, a fierce king, *The Scarecrow of Oz*.

Lord Pinkerbloo, pink or blue, *Rinkitink in Oz*.

Rak, a giant creature who breathes out smoke (kar or "car" spelled backward), *Tik-Tok of Oz*.

Tallydab, Tellydeb, Tillydib, Tollydob, Tullydub, five counselors to Queen Zixi (each of the five vowels appears twice in each name), *Queen Zixi of Ix*.

Tietjamus Toips, a musician ("Tietj" begins the last name of Paul Tietjens, Baum's friend who wrote the music for the stage version of *The Wizard of Oz*. Could the remainder of the name refer to pajama tops?), *John Dough and the Cherub*.

Tip, Ozma in enchanted form as a boy. When she was later kidnapped by Ugu, he transformed her into a peach pit (pit is tip spelled backward), *The Marvelous Land of Oz*.

Ugu, an evil shoemaker. Ugu appears in the third column of letters containing Baum's wife's full name; coincidence?

MAUD
GAGE
BAUM

Woot, the protagonist of *The Tin Woodman of Oz*. The book's title has the acronym TWOO. Move T from front to back to get WOOT, *The Tin Woodman of Oz*.

Will-Takum, a thief who will take whatever he wants, *The Enchanted Isle of Yew*.

Queen Zixi of Ix, shift the Hebrew letters for Jehovah—YHWH—forward one step and you get Zixi; shift XI back nine steps to get OZ, *Queen Zixi of Ix*.

Oz is divided into four regions: Gillikin in the north, Winkie in the west, Munchkin in the east, and Quadling in the south. The Gillikin region, where purple is the dominant color, may have been named after the purple blossoms of the gillyflower, a plant that flourishes in upper New York where Baum spent his boyhood. I have also speculated that Munchkin may refer to the munching of breakfast after the sun rises in the east. The color of this region is blue, perhaps a reference to the sky turning blue early in the morning. Winkie may refer to the winking of sleepy eyes when the sun sets in the west. The Winkie color of yellow could suggest the gold of sunsets. Quadling clearly stands for the fourth (quad) region of Oz.

The most puzzling of Baum's proper names is Tititi-Hoochoo, the handsome black-eyed ruler of a land at the other end of a tube that goes through the earth. It has been suggested that the name sounds like a sneeze, but Baum may have had something else in mind. My conjecture is that Baum originally intended the ruler to be a beautiful dark-eyed dancer, combining a reference to her breasts with the erotic dance called the hootchie-cootchie. The book's publisher forced him to change the ruler to a man to avoid titillating child readers.

POSTSCRIPT

After each entry in this chapter I give the title of the Baum book in which the word first appears. I considered only

Baum's Oz books and his major fantasies. I did not include such lesser fantasies as *Animal Fairy Tales, Babes in Birdland, The Wogglebug Book*, and numerous short stories that appeared only in periodicals.

I was not, of course, serious in mentioning the word shifts of OZ, or the word shifts of Queen Zixi. These are obviously coincidences. The backward shift of OZ to NY was discovered by Mary Scott. I confess responsibility for the forward shift to PA. Michael Hearn was the first to speculate that Dot and Tot refer to Dorothy and Toto. Fred Meyer discovered Ugu's name concealed in Maud Gage Baum. Jack Snow conjectured the origin of Woot's name.

Word play was something Baum and Lewis Carroll had in common, especially a fondness for puns. Some of Baum's puns are easily missed, such as when General Jinjur tells the Guardian of the Emerald City gates that she and her girls are "revolting." The soldier replies, "You don't look it." Two chapters in *The Emerald City* swarm with puns — the chapter on Utensia and the one on Bunbury. A raft of puns pepper *John Dough and the Cherub*.

Harry Tobias, in three articles in *The Baum Bugle* (Christmas 1972, autumn 1974, and spring 1978) was the first to search for Baum's linguistic play. I took up where he left off by concentrating on proper names. The interpretations fall into three classes:

1. Meanings Baum clearly intended.
2. Meanings he may or may not have intended.
3. Meanings he did not intend, but which can be amusing nonetheless, such as the word shifts cited above, and Fred Meyer's quip that Dr. Nikidik (*Land of Oz*) often performed circumcisions.

Meyer, looking over an early draft of my list, added the following possibilities:

1. In the early years of the twentieth century Michigan had a governor named Frank Pingree. Since Baum spent summers in Michigan, he surely knew of Pingree, and may have played on his name in naming the Island of Pingaree (*Rinkitink.*).

2. Pastoria, Ozma's father and former king of Oz (*Land of Oz*) could have been named for Francis Pastorius, the first German immigrant to come to the United States.
3. Evoldo, king of Ev, spelled backward is "odd love," and pronounced right you get "evil do." Errob and Evedna, prince and princess of Ev, reflect the names of Baum's second son, Rob, and his wife Edna (*Ozma*).

Ruth Berman, who sent lots of good suggestions, thinks the giant Yoop (*Patchwork Girl*) may have gotten his name from the Yoopers, a slang term for residents of upper Michigan noted for their strange way of speaking and their odd customs.

Coo-ee-oo is an evil queen (*Glinda*). According to Eric Partridge's *Dictionary of Slang*, "Cooee" is an Australian Black's signal cry, adopted by the colonists and later heard often on London streets. Gore Vidal, in an essay on the Oz books, recalls it as a call of hog callers, and Fred Meyer heard it on farms as a call of cows.

When Baum named Ojo the Unlucky (*Patchwork Girl*), did he know that ojo is Spanish for "eye"?

The High Coco-Lorum, king of the Twists (*Lost Princess*) is surely named after the slang word cockalorum. Partridge defines this an overconfident little man. The exclamation "high cockalorum," he adds, was popular around 1800.

Correspondent Karl Hill startled me with a letter saying that the German word *menschchen* means little people! Baum, of German ancestry, must have known the word and applied it to the little people of Munchkin land.

There is a King Scowleyow in Baum's *Magical Monarch of Mo*. Jim Haff located a poem titled "Snarleyow," in Kipling's *Barrack Room Ballads*, about a military horse of that name, and Ruth Berman turned up a novel by Captain Frederick Marryat, *Snarleyyow* (1839) about a dog of the same name.

Dave Morice wrote in a letter that Inga is also concealed in my own name: MartIN GArdner.

I have compiled an extremely long alphabetized list of all the unusual nouns in Baum's fantasies, along with their possible meanings, including words for which I have no meanings. I may publish it someday—it would make a small book. An invaluable reference has been Peter B. Clarke's *Who's Who, What's What, and Where's Where in Oz: An Index to the People, Creatures, Things, and Places in Oz in the First Forty Books* (1994).

HUGO GERNSBACK

When I was a preteen in Tulsa, the great delight of my boyhood was a now almost forgotten magazine called *Science and Invention*, edited and published by Hugo Gernsback. I often wonder how many elderly scientists and science writers today owe their initial enthusiasm for science to this wonderful, outlandish periodical.

Gernsback is best known today as the Father of American Science Fiction. At annual science-fiction conventions awards for the year's best writing and illustrating are models of a space rocket called a "Hugo." But Gernsback was much more than a pioneer publisher of science-fiction magazines. He was one of the nation's great science popularizers.

Born in Luxembourg in 1884, the son of Mauritz Gernsbacher, a Jewish wine wholesaler, young Hugo attended two technical schools, one in Luxembourg, the other in Germany, before coming to New York City in 1904. He was fired with a strong sense of how rapidly the world was be-

Hugo Gernsback

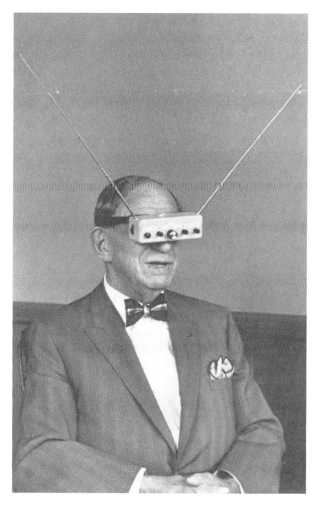

Hugo Gernsback, wearing goggles that allow stereoscopic viewing of two miniaturized television screens.

ing transformed by electrical technology. In 1905 his Electro Importing Company was America's first radio mail-order firm. It also sold radios and electrical equipment from its store at 69 West Broadway. Batteries were supplied by Gernsback's GeeDee Dry Battery Company. In 1906 *Scientific American* advertised, and department stores carried, Gernsback's Telimco Wireless that rang bells and transmitted Morse code.

The year 1908 was the beginning of Gernsback's long

career as a magazine publisher. His first periodical, *Modern Electrics*, in 1908 was the nation's first magazine devoted mainly to radionics. After selling the magazine in 1913, Gernsback began a new monthly called *The Electrical Experimenter*. In 1920 the name was changed to *Science and Invention*. Alexander Graham Bell and Thomas Edison were among famous inventors who sent Gernsback congratulations on the name change. When *Science and Invention* became *Science and Mechanics* in 1932, Gernsback continued as editor until four years later when he sold the magazine to another publisher.

What Gernsback originally called "scientifiction" appeared regularly in *Modern Electrics*, *The Electrical Experimenter*, and another of his periodicals called *Radio News*. To meet increasing reader demands, *Science and Invention* began including one or more such tales in every issue. Among them, forty stories about the amazing inventions of Dr. Hackensaw were written by Clement Fezandie, a New York City science teacher. Although Fezandie's writing was undistinguished, his stories introduced a raft of themes that would soon be taken up by more skilled authors. *Science and Invention* also featured novel-length serials by such early science-fiction writers as Ray Cummings and A. Merritt.

In 1926 Gernsback started *Amazing Stories*, the world's first magazine devoted exclusively to science fiction. I was a charter subscriber; would that I had preserved the first year's issues! When Gernsback's Experimenter Publishing Company declared bankruptcy in 1929, he lost *Amazing Stories*, but at once bounced back with *Science Wonder Stories*. This was soon followed by four other science-fiction pulps, and in 1953 by a slick-paper monthly, *Science Fiction Plus*. An earlier Gernsback periodical, *Superworld Comics*, the world's first science-fiction comic book, survived for three issues.

Today's critics are harsh on Gernsback for overplaying hard science to the neglect of style, plot, and character. He in turn became dismayed by a trend toward abandoning

Hugo Gernsback at 75

facts for fantasy. Modern authors, he wrote in *Science Wonder Stories*, "do not hesitate to throw scientific plausibility overboard, and embark on a policy of what I might call scientific magic . . . science that is neither plausible nor possible. Indeed, it overlaps the fairy tales, and often goes the fairy tale one better."

Aside from his science-fiction magazines, Gernsback published more than forty other periodicals. They had such titles as *Your Body*, *Tid Bits*, *New Ideas for Everybody*, *Cooko Nuts*, *French Humor*, *Radio-Craft*, *Radio News*, *Radio Review*, *Radio-Electronics*, *Television News*, *Short-Wave Craft*, *Milady*, *Scientific Detective Monthly*, and *Sexology*. Except for *Sexology*, which had a long run, most of these magazines were short-lived. A notable failure was *Technocracy Review*. It featured essays pro and con about the technocratic movement which expected science experts to become the future managers of civilization. Gernsback, always fond of inventing names, thought "scientocracy" was a better term.

Gernsback also issued a variety of one-shot specials such as a selection of puzzles by Sam Loyd, and a collection of magic tricks by Joseph Dunninger. Each December he distributed to friends and associates "Christmas card" booklets that parodied popular magazines. They had such titles as *Harpy's Bizarre*, *Like*, *Newspeek*, *Quip*, *Popular Nekanics Gagazine*, *The Saturday Evening Host*, *Jolliers Digest of Digests*, and *Tame*, a spoof on *Time* dated 2045.

In 1928 Gernsback offered $500 in prizes for the best answers to what is wrong with the magazine's cover. The December issue reported that readers found 48 mistakes.

One of Gernsback's pseudonyms for an occasional hoax article was Grego Banshuck, an anagram of his name.

Science and Invention was far and away Gernsback's most successful magazine. Like its predecessor, it was an exciting mix, rich with photographs and drawings, of science news, departments for amateur experimenters, and wild speculations about future technology, all infused with Gernsback's love of science and awe over how it was altering history. For five years its striking covers, by Austrian-born Frank Rudolph Paul, were printed on gold colored paper to symbolize the golden age of science. Paul also did the covers and much of the inside black and white illustrations for Gernsback's science-fiction pulps. Arthur Clarke has called him the "undisputed king of the pulp artists."

In September 1927 Gernsback's monthly editorial in *Science and Invention*, headed "Twenty Years Hence," predicted that planes would go from New York to Paris in twelve hours, television would revolutionize our way of life as much as the telephone, the average human life span would increase to 70, and all homes and buildings would be air conditioned. There were as many misses. Gernsback believed cancer would be eliminated by 1947, electrical power would be transmitted by wireless, and airplanes would settle on one spot by descending in a slow spiral.

It would take many pages merely to list the topics covered in *Science and Invention.* Here are some. Color movies, talking movies, and ways to add a third dimension to the screen. Flame throwers, fax machines. How dinosaur models were constructed for films such as Conan Doyle's *Lost World.* Can animals think? Implanting hair on bald heads. Can freezing prolong life? Will rocket missiles be fired in the next world war? How Egyptians built their pyramids. Is the universe expanding with increasing speed or slowing down? Will melting glaciers flood New York? Will helicopters help build skyscrapers? How to send pictures by wireless. A voice recorder for airplanes. Many articles explained relativity theory and evolution. A young

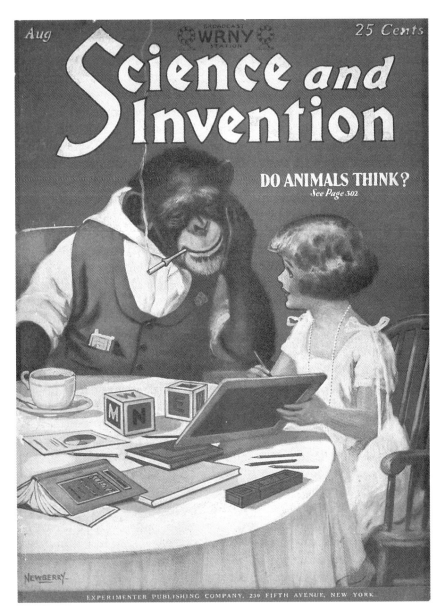

Raymond Ditmars, a curator at New York City's zoo, argues in 1928 that many animals do indeed think. A second article prints some of the replies to a questionaire sent by the magazine to hundreds of scientists asking if they believe animals think. The majority said yes.

Donald Menzel, later to become director of Harvard's observatory, contributed articles on astronomy. In editorials and articles Gernsback predicted that soon homes would be heated by solar energy. Gravity waves, moving at the speed of light, would someday be detected. In 1916 he foresaw the time when a president's speech would be broadcast to "all the people." That same year he wrote: "In less than fifteen years every automobile . . . will carry its small radiophone." Paul's cover for April 1924 showed a man and woman kissing, with electrodes attached to parts of their body to make measurements. It illustrated Gernsback's article, "Scientific Mating."

I first learned about Klein bottles from a picture in the June 1924 issue. This topological curiosity is a closed, edgeless surface, like that of a sphere, except that, like a Moebius strip, it has only one side. As the text explained, if you punch a hole anywhere on the bottle's surface, and insert a string, the string's ends can always be tied.

Gernsback took for granted that it was only a matter of time before physicists would split the atom. "The atomic energy locked up in a one-cent piece," he wrote in March 1918, "is sufficient to lift up the Woolworth building. . . . The new energy is coming as surely as the sun will rise tomorrow."

In a June 1921 editorial he likened today's physicists to ancient savages who watched forests burn but had not yet learned how to make fire. "We know today . . . that if we could harness the inherent atomic energy of one gram of iron it would equal the explosive force of fifty tons of dynamite. But as yet we are savages, and it is Nature's wisdom not to let us play with super savages, and it is Nature's wisdom not to let us play with superforces for a while at least. Which may be for the best—because we probably would use the new force to kill each other in the next Great War."

Regular sections of *Science and Invention* were devoted to the latest science news, curious patents, and inventions most needed. There were motor hints for car

Long before Masters and Johnson, Hugo Gernsback writes about ways science can test whether a man and woman are sexually compatible.

This 1928 issue gave detailed instructions on how to build a television receiver using a rotating disk with a spiral of holes for scanning the tiny screen. Gernsbach was broadcasting images daily from WRNY, his Manhattan radio station. It was the first regular broadcasting of television in America.

owners, and experiments in physics and chemistry to perform at home. Readers were told how to make radios, telescopes, phonographs, air conditioners, theramins, and all
sorts of electrical devices. One issue even provided plans
for building an airplane!

In a July 1915 editorial in *The Electrical Experimenter*
Gernsback anticipated sonar as a way to locate enemy
submarines:

> We can imagine an apparatus, say at the bow of the ship,
> sending out waves below the water while a suitable
> detector at the stern, also below water, is used to register
> the "echo" and its intensity. The original waves striking
> the submarine will be reflected and bent back. The inten-
> sity of these reflected waves could be made to read off on
> a direct recording scale, giving the distance of the sub-
> marine in miles. Recent researches show the possibility
> of sending . . . waves below water, so we may be sure that
> the interesting problem of locating submarines will not
> remain unsolved for any great length of time.

Gernsback was not the first to use the term television—
France beat him to it—but he was the first to introduce it to
American readers. In 1928 his radio station WRNY actually
began broadcasting television images during the first five
minutes of every hour. It was the nation's first such daily
broadcasts. Of course viewers had to build their own
receivers, as described in *Science and Invention*'s November
1928 issue. The crude method used a disk two feet across
with 48 holes arranged in a spiral and spinning at 540 rota-
tions per minute. One disk scanned the studio's image while
a duplicate disk, synchronized with the studio's, scanned
the receiving screen. No sound was transmitted and the
screen was a square one-and-a-half inches on the side. The
New York Times began listing these broadcasts on August 1,
1928, and articles about television began to appear in almost
all later issues of *Science and Invention*, as well as in Gernsback's magazines devoted to radionics.

Hugo Gernsback

Popular Television

Built and Described by the Staff

How To Build The S & I TELEVISION RECEIVER

A slight adjustment of the rheostats and the picture comes in clearly. This photo shows a complete television receiver connected to an ordinary radio set. The picture is seen in the cone.

THE front cover illustration shows the simple television receiver designed and built by the editorial staff. The accompanying photographs and drawings show the appearance and the construction details of the television receiver, the apparatus pictured having, of course, to be connected to the output of a suitable radio receiving set. The ideal set for receiving television images from WRNY or other stations, is, for the broadcast wavelength of 326 meters, one comprising two or three stages of tuned radio frequency, a detector and at least three stages of resistance-coupled amplification. When a resistance-coupled amplifier is used, it will be found best to use above 250 volts at least on the last stage from either storage or dry "B" batteries. A good "B" eliminator may be used, but a special filter is usually necessary, to prevent "motor-boating" with a resistance-coupled amplifier.

PROPER MOTOR FIRST ESSENTIAL

THE first requisite for building this television receiver is a good 16-inch fan motor. If the television disc to be used (it should have 48 holes for reception from WRNY and 3XK; also 1XAY and WLEX of Boston; and 24 holes for reception from WGY, 2XAD, and 2XAF, G. E. Co., Schenectady), is quite light, a 12-inch fan motor may do the work. If you have direct current in your laboratory or other location where the apparatus is to be operated, then you will have no trouble in controlling the speed of the motor down to the 450 r.p.m. required for WRNY reception or the 900

A Television Receiver of Simple Design, Built Around an Ordinary 16-inch Electric Fan Motor

r.p.m. required for reception from the other stations broadcasting television.

If you have to select or use an alternating current fan motor, then you will have to

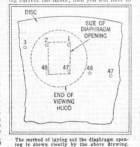

The method of laying out the diaphragm opening is shown clearly by the above drawing.

find out whether the motor can be slowed down to a steady speed of 450 r.p.m. If the A.C. motor happens to be of the type that has throw-out contact brushes, which open the starting winding after the motor has attained fairly high speed, you will probably find this sort of motor unfit for television purposes. If the motor is of the universal A.C.-D.C. type, with commutator and brushes, the armature being connected in series with the field, then you will find that this motor can be regulated as to speed very nicely by means of the series resistances shown in the accompany diagram. We strongly recommend a universal type motor if you are going to purchase one, as these have been found to regulate well with regard to the speed.

MOUNTING THE DISC

THE disc used in the television receiver here illustrated was a 48-hole 16-inch diameter bakelite disc of standard manufacture. This disc may be mounted and secured on a regular bushing provided with lock nuts supplied by the people who make the disc. In the present case, however, the perforated disc was mounted on the brass spider and hub which had originally carried the fan blades. The blades were removed from the legs of the spider and these were then flattened out in a vise and checked up on a lathe for alignment. A light cut may be taken across the face of the spider legs in the lathe, if one is handy. By drilling holes through the bakelite disc, it is readily secured to the spider by machine

The first page of an article in November 1928 explaining how to build a television set for receiving the images sent from Gernsback's radio station WRNY.

115

Endless prophetic misses were as frequent in *Science and Invention* as its spectacular successes. Wild trains, cars, ships, and planes were described. A ship pictured in December 1923 was made of five sections hinged together so that when waves bounced them up and down its gears would generate power that propelled the ship. Trains were shown sliding on water film, or with steel balls substituted for wheels. There were pictures of Martians and maps of Mars' canals. Was Jupiter's Red Spot a moon in the making? Are plants growing on the Moon? Will robots someday serve as policemen? Can gravity be nullified? Can people learn while asleep? How did Noah's Ark manage to house all the animals, and what was the source of the flood waters? Is there a planet inside Mercury's orbit? Will cars some day run on water?

Some other topics: ways to extract gold from the sea, how to obtain energy from sea waves and tides, a mind-reading dog, electrically powered ice skates, the lost continents of Atlantis and Lemuria, and an underground tube for rapid transportation between San Francisco and New York. My vote for the magazine's nuttiest notion was a giant centrifuge, shown on the cover, with elderly patients in beds around the rim. The centrifugal force generated was supposed to counter the aging effects of gravity!

More amusing misses: How to photograph the human aura. Can persons be rendered invisible by altering their atoms? A boomerang metal shield curved to send bullets back to an enemy. Will we ever evolve an ability to visualize the fourth dimension? Can thoughts be recorded? In January and February 1919 Nicola Tesla, Gernsback's good friend, contributed two foolish articles in which he struggled to prove that the Moon does not rotate on its axis.

New developments in all the sciences were regularly reported in *Science and Invention*. At the same time Gernsback delighted in debunking bogus science. Scores of articles exposed such follies as perpetual motion, medical quackery, fake Spiritualist mediums, and the idiocy of

In 1926 field editor Joseph Kraus blasts astrology with unassailable arguments. The magazine offers $5,000 to any astrologer who can foretell three major events of such a nature that he has no control over the outcome, and $1,000 for three accurate detailed horoscopes on the lives of three persons when given only their initials and birth dates.

astrology. Common science mistakes were regularly countered, such as the belief that stars can be seen in daytime from the bottom of deep wells.

Magic was another favorite topic. Dozens of articles gave instructions for tricks with common objects. Ten were written by Walter Gibson, later to become famous for creating the radio and pulp magazine hero, the Shadow, and for his many books on conjuring. Magician Joseph Dunninger supplied a monthly feature on magic. Stage illusions were explained. Gaffed carnival games and methods of cheating by gamblers were revealed. Articles dealt with the secrets of special effects in popular films.

Gernsback was fond of offering cash prizes to winners in more than twenty contests based on what can be done with old-razor blades, bottles, phonograph needles, gold paper, clock springs, inner tubes, wood matches, wires, pipes, and so on. Fifty thousand dollars was offered for a workable perpetual motion machine, and $21,000 for proof that we can communicate with the dead. One unusual contest involved finding useful combinations of things such as dental floss at the end of a toothbrush, or a nail file on a scissors blade.

For years *Science and Invention* ran a humor page offering a dollar for every science joke printed and three dollars for the best joke of the month. Two examples:

Here lies William Johnson.
Now he is no more.
What he thought was H_2O
Was H_2SO_4

Teacher: What is a polygon?
Student: A dead parrot.

Gernsback did his best to write his own science fiction. Of his two novels the most successful in sales was *Ralph 124C 41+: A Romance of the Year 2660*. First serialized in *Modern Electrics* in 1911, it was later published in both hard and soft covers, and translated into French, German, and Russian. Its inept writing and preposterous plot have earned it the distinction of being the worst science-fiction novel ever published.

The book's hero is Ralph, a world famous scientist who is one of ten living persons allowed to add a plus sign after his number. A faulty phone conversation puts him in touch with Alice 212B 423, a dark-eyed Swiss girl. While they converse on the telephot—each can hear and see the other —Alice realizes an avalanche is about to demolish her home. Ralph moves quickly to the roof of his Manhattan laboratory where he directs a powerful energy beam that reduces the rushing snow to harmless hot water.

Ralph and Alice are soon in love. He has two rivals, the evil Fernand, and Llysonorh, a tall Martian humanoid. Fernand kidnaps Alice by first rendering her invisible. Ralph rescues her only to have Fernand kidnap her a second time and carry her off in his spaceship. Ralph gives chase in his own space-ship, powered by inertial drives. He links up with Fernand's craft, goes inside, and finds him unconscious. Fernand's ship had been boarded by Llysanorh, also madly in love with Alice. He has kidnapped her for the third time and is taking her to Mars. Unable to catch up with Llysonorh's craft, Ralph trans-forms his own ship into a comet on a collision course with the red planet. He reasons that the Martian government will send Llysonorh out to alter the fake comet's path.

Sure enough, the Martian, with poor Alice still on board, heads for the "comet." Ralph links his ship to the Martian's craft. He kills Llysonorh, but Alice has been stabbed by the Martian and has bled to death. Ralph refrig-erates her corpse, takes it back to his laboratory, and with the help of 16K 5+, the world's greatest surgeon, restores Alice to life. The novel ends with a dreadful series of puns. While Ralph twines a finger around a tendril of Alice's hair, she whispers to him that she has discovered the secret meaning of his number name. It is "one to foresee for one."

The novel is still read today—its first edition is a rare

collector's item—mainly for its amazing anticipations. In chapter 11, for example, there is a clear description, with a schematic illustration, of radar. Radio waves beamed toward a spaceship will bounce back, then "from the intensity and elapsed time of the reflected impulses, the distance between the earth and the flyer can then be accurately calculated." In 1927 Gernsback proposed bouncing radar waves off the moon, a feat that did not occur until 1946.

There are many other good predictions in the novel: juke boxes, aluminum foil, plastics, TV screens attached to phones, synthetic fibers, flourescent lights, sky writing, night baseball, space sickness, microfilm, stainless steel, musak, tape recorders, cryogenics, liquid fertilizer, color photos in newspapers, and many others.

Although these hits are balanced by wide misses, because the novel is set in 2660, who can be sure they will not eventually materialize? They include three-dimensional television, cableless elevators operated by electromagnetism, life spans of 149 years, crops electrified and warmed by Earth's interior heat, cities with total weather control that keeps them in perpetual sunshine. Rain and snow are permitted occasionally in early morning. A harmless gas stimulates appetite in "scientificafes." Tubes to the mouth send liquid food to the stomach. It took many decades, Gernsback writes, for people to overcome the habit of mastication.

A "language rectifier" instantly translates languages on the telephot. A "memograph" types out thoughts. The "hypnoscope" sends entire books to a sleeper who remembers them next day. At 5 A.M. newspapers are transmitted to sleeping subscribers. A huge vacation city, with shops and playgrounds, is under a glass dome, suspended above the earth by an antigravity screen.

Plot summaries of Gernsback's other science fiction can be found in Everett Bleiler's massive reference *Science-Fiction: The Gernsback Years* (1998). A 1918 story, "The Magnetic Storm," anticipated a memorable scene in the film *The Day the Earth Stood Still.* Gort, a powerful robot, stops the movement of all vehicles around the world. Gernsback's story tells how a scientist ends a great war by surrounding Germany with a magnetic field that paralyzes all machines that use electricity.

Hugo Gernsback was a slim, dark haired man with enormous energy, a canny business sense, and a passionate love of science and technology. Always nattily dressed, in later years he used a monocle for reading restaurant menus. He spoke French as well as English, both with a thick German accent.

"I believe," Gernsback once wrote, "that the average type of 'nut' inventor is one of the greatest assets to civilization." A nut inventor himself, he is said to have obtained some eighty patents, though I am not aware of any that were profitable. They included an electric comb and hairbrush, an apparatus for launching airplanes, a luminous mirror, an electric fountain, and a device for hearing through the teeth. An "isolator" helmet, with an attached air hose for breathing, allowed one to think and work without being distracted by noises. Gernsback also wrote and published books and service manuals with such titles as *Radio for All*, *The Wireless Telephone*, *Wireless Hookups*, and *Television Repair Techniques*. In the 1960s he issued *Forecast*, an annual Christmas booklet in which he predicted that frozen corpses would be shot into space, material objects would be electronically transmitted ("Beam me up, Scotty!"), and stereoscopic goggles containing miniaturized television screens for each eye.

Gernsback died in 1967, survived by his third wife and three of his four children. His only son Harvey took over Gernsback Publications and later sold it to Larry Steckler who runs the firm from Farmingdale, New York. It currently publishes two magazines, *Popular Electronics* and *Electronics Now*. The company has a Web site at www.gernsback.com.

Gernsback's nephew Patrick G. Merchant, a computer specialist who runs Information Technologies Group, in

Johnston, Pennsylvania, has a website, "Hugo Gernsback's Forecast," honoring his uncle: http://www.twd.net/ird/.

Arthur C. Clarke's dedication in *Profiles of the Future* is "to Hugo Gernsback, who thought of everything." Well, not quite.

Like other science-fiction writers and futurists, including H. G. Wells, Gernsback had no inkling of the computer revolution. This is not to fault him. Who among us today can even imagine the marvels of technology the new century will bring?

FURTHER READING

"Gernsback, the Amazing," *Time*, 3 January 1944, pp. 41–42.

Sam Moskowitz, "Hugo Gernsback," chapter 14 in *Explorers of the Infinite: Shapers of Science Fiction* (World Publishing, 1963).

Paul O'Neil, "Barnum of the Space Age," *Life*, 15 July 1963, pp. 62–68.

Mark Siegel, *Hugo Gernsback, Father of Modern Science Fiction* (Borgo Press, 1988).

Daniel Stashower, "A Dreamer Who Made Us Fall in Love with the Future," *Smithsonian* (August 1990): 45–54.

BOOK REVIEWS

CARL SAGAN'S DEMON-HAUNTED WORLD

M ost scientists prefer to ignore cranks. Trying to enlighten them or their followers, scientists say, is like trying to write on water. Besides, scientists believe they have more important things to do. Carl Sagan, Cornell's distinguished astronomer, sees it differently. He has become the fugleman of a growing band of experts now convinced that the rising flood of superstition and pseudoscience is too damaging to society to be ignored. How can political leaders vote intelligently on proposals to control nuclear weapons, to fund basic research, or to reverse environmental degradation if they and their constituents can't distinguish good science from bad?

Although this "dumbing down," as Sagan terms it, is global, it is on our shores that the virus of scientific illiteracy has become unusually toxic. As Sagan tells us, polls

This review first appeared in the *Washington Post*'s *Book World*, March 17, 1996.

show that 57 percent of Americans don't know that dinosaurs died out before humans evolved. Indeed almost half the populace thinks the earth was created about 10,000 years ago, and that the dinosaurs perished in Noah's Flood. (They were too big to go on the Ark.) Most citizens don't know that the moon and stars rise and set like the sun or that electrons are smaller than atoms.

The *Demon-Haunted World* (Random House, 1995) is a stirring defense of informed rationality. It is also a powerful indictment of today's miserable science teaching, the upsurge of Protestant fundamentalism, and the roles of greedy book publishers, abetted by the print and electronic media, in accclerating America's dumbing down,

Few believers in crank science escape Sagan's withering blasts. Let us name a few. Christian Scientists who would rather let their children die than give them insulin or antibiotics. Creationists who think God made Eve out of Adam's rib. Fools who ignore sound medical treatment to be "operated" on by Philippine psychic "surgeons." Ignoramuses who waste money on homeopathic drugs diluted until no molecule of the original substance remains. Does the water "remember" the drug's properties?

Sagan devotes hilarious paragraphs to his friend John Mack, Harvard's peculiar psychiatrist who treats patients he firmly believes have been abducted and often abused by aliens on UFOs. An embarrassed Harvard can't fire him because he has tenure. There are ten times more astrologers in the United States, Sagan reminds us, than astronomers. Are astrology books and newspaper columns as harmless as fortune cookies? Recall the sad spectacle of our past president and first lady scheduling important meetings on dates set by their West Coast star gazer. Orange County, California, declared bankruptcy last year because its treasurer made stupid investments recommended by a mail-order astrologer.

Other follies covered by Sagan include the supposed stone face on Mars; the ridiculous crop-circle hoaxes; J. Z.

Knight's profitable channeling of Ramtha, a 35,000-year-old CroMagnon warrior who lived on the newly sunken Atlantis; the fake shroud of Turin; bleeding and weeping statues of the Virgin Mary; Big Foot and the Loch Ness Monster; spontaneous human combustion; poltergeists; perpetual motion machines; ESP and psychokinesis; dowsing; auras; Scientology; Lysenkoism; pyramid and crystal power; the Yogic flying of Transcendental Meditators; faith healing; facilitated communication with autistic children—the litany is endless.

The most tragic result of scientific ignorance, skillfully detailed by Sagan, is the current mania for sentencing innocent parents and preschool teachers to prison, sometimes for life, solely on the basis of false memories of childhood sexual abuse fabricated by incompetent, self-deluded therapists and social workers. The false memories often involve horrible cannibalistic rituals in Satanic cults that the FBI has declared nonexistent. Another prime example of how scientific ignorance can thwart justice, too recent to be in Sagan's book, was the inability of O. J. Simpson's jurors to comprehend the accuracy of DNA blood testing.

Like all of Sagan's books, *The Demon-Haunted World* is rich in surprising information and beautiful writing. His colorful attacks on crazy science are accompanied by happy memories of his childhood in Brooklyn, his early enthusiasms, his passionate love of science and the influence of his parents, his college mentors and Ann Druyan, his wife. (I, too, am handsomely acknowledged, though I don't know Sagan personally and didn't provide any help in the writing of his book.) Philosophical questions are not ignored. Sagan has no use for sociologists who pretend that scientists invent rather than discover laws of nature. Science, he insists, is much closer to mathematics than to cultural differences and changes of fashion. Two plus two is four in all possible worlds. The laws of quantum mechanics are the same in China as in America.

Although science is always corrigible—Sagan even lists

several of his own past mistakes—who can doubt that it moves inexorably closer to understanding how the universe behaves? Reality is out there, independent of our little minds.

Sagan's funniest chapter contains his account of a notorious swindle perpetrated on Australians in 1988 by magician James Randi, the world's top baloney-buster. Randi cunningly orchestrated the appearances in Sidney of a young man who claimed to be a powerful psychic and New Age guru called Carlos. Gullible Australians tumbled completely for the hoax, and were fit to be tied when it was revealed. To promote his psychic, Randi actually wrote and distributed a booklet titled *The Teachings of Carlos*. Sagan quotes some typical passages:

> Why are we here? . . . Who can say what is the one answer. There are many answers to any question, and all the answers are right answers. It is so. Do you see? . . . Of doubters, I can only say this: let them take from the matter just what they wish. They end up with nothing—a handful of space, perhaps. And what does the believer have? EVERYTHING! All questions are answered, since all and any answers are correct answers. And the answers are right! Argue that, doubter.

Sagan is not unaware that the outrageous claims of religious superstition and junk science arouse great awe and wonder. It is spine-tingling to suppose that we are surrounded by invisible angels and demons. How thrilling to believe that psychics can bend spoons with their minds, predict the future, and view scenes thousands of miles away. Such ersatz wonders, however, pale beside those of authentic science-wonders that glow throughout all of Sagan's marvelous, wonder-saturated books. Here is a small sample, a footnote in *The Demon-Haunted World*:

> Except for hydrogen, all the atoms that make each of us up—the iron in our blood, the calcium in our bones, the

carbon in our brains—were manufactured in red giant stars thousands of light years away in space and billions of years ago in time. We are, as I like to say, starstuff.

ADDENDUM

The letters that follow, along with my reply, appeared in *Book World* (May 12, 1996):

We tend to venerate knowledge so much in this country that we listen to scientists even when they make statements for outside their area of expertise. I was dismayed to read Martin Gardner, who is presumably echoing the ideas in Carl Sagan's book *The Demon-Haunted World* when he disparages "ignoramuses who waste money on homeopathic drugs diluted until no molecule of the original substance remains. Does water 'remember' the drug's properties?"

The answer to his rhetorical question is "perhaps." Something is happening. I am a homeopath and I have used homeopathic medicine to treat illness in myself and I have seen it work in others, including children, infants, and even animals who are presumably not capable of being influenced by the placebo effect. Homeopathy has been used successfully to treat illness for over 200 years. One of the best-selling flu remedies on the market is Oscillococinum, which is homeopathic and works at a potency below Avogadro's limit. The book *Homeopathy, Medicine for the Twenty-first Century* details some of the controlled experiments conducted on homeopathic medicines.

Homeopathic medicines are produced by diluting the basic substance, an herb for the sake of argument, one to a hundred times and vigorously shaking the solution and repeating the dilution and shaking (or succussion) a specific number of times. These solutions work; that is, they heal illness, below the theoretical limit where little or no molecules should exist. Yes, if one merely dilutes a herbal substance too far without either the succussion or the serial dilution, the solution loses its effectiveness at a certain point as one would expect from Avogadro, high school chemistry, and nineteenth-century physics. Why homeopathic medicines continue to be effective in healing illness when diluted and succussed below this theoretical limit is not understood. Carl Sagan (and Martin Gardner) should add to their commentaries that they have no experience with the efficacy of homeopathic medicines. They have done no experiments to confirm or deny the effectiveness of homeopathic medicines. The best Sagan can say is that theoretically, based upon the current "scientific" model, some homeopathic medicines should not work, even though they do.

The story comes to mind of the aeronautical engineer who looked at the body mass and wing span of a bumblebee and concluded that it was impossible for it to fly, or the learned men of Galileo's time who refused to look through the newly invented telescope at the moons of Jupiter because they violated the "scientific knowledge" of the day—that everything revolved around the earth. When Carl Sagan talks from this basis, he is being just as foolish.

One day, as science advances, open minds will discover the exact mechanism by which homeopathic medicines work. Until then, these medicines will continue to be used in the same way that doctors have used aspirin for nearly 100 years and still to this day are not certain how it and many other prescription medicines work.

Roger Ashford
Annandale

In his review of Carl Sagan's new book, *The Demon-Haunted World*, Martin Gardner, a well-known science writer himself, rightly applauds Sagan's "stirring defense of informed rationality," his call for better science teaching in the schools, and his concern about "America's dumbing down."

However, Gardner repeats Sagan's startling charge that "Christian Scientists . . . would rather let their children die than give them insulin or antibiotics." It is utterly untrue that Christian Scientist parents, or those engaged in the healing practice of this respected, world-wide religion, would rather have children die than be treated with medicine. Christian Scientists, like all loving parents, consider their children's health of utmost importance. The reason they turn to scientific prayer is because they have actually experienced its healing efficacy in their own and their children's lives, and have come to rely on it as naturally as do others who rely on conventional medicine. As far as the church is concerned, they are free to choose whatever form of treatment they deem best, without fear of censure.

Much of what Sagan and Gardner deride as "unscientific" simply manifests dissatisfaction with the gross materialism of modern science and medicine. There is a yearning today for something beyond the strictures of matter, for something more of the mind and spirit, for a greater understanding of God and of His actual presence and power in our lives.

Nor is this reaching beyond matter limited to what they may think of as the gullible fringes of society. Last December in Boston, over 900 medical and other professionals attended a four-day symposium on "Spirituality and Healing In Medicine." Harvard Medical School and the Mind/Body Medical Institute of Deaconess Hospital cosponsored the meeting. It included discussions by members of the medical and religious communities on the beneficial effects of prayer.

Sagan and Gardner might contribute even more to enlightened thinking than they already do by recognizing the limits of matter and by acknowledging the undeniable evidence of the power of prayer and spirituality in human life.

Ralph E. Burr
Christian Science Committee on
Publication for the District of Columbia

The following is what Carl Sagan wrote about Christian Science in his book *The Demon-Haunted World*: "Faith and prayer may be able to relieve some symptoms of disease and their treatments ease the suffering of the afflicted, and even prolong lives a little. In assessing the religion called Christian Science, Mark Twain—its severest critic at that time—nevertheless allowed that *the bodies and lives it had made whole by the power of suggestion more than compensated for those it had killed withholding medical treatment in favor of prayer"* (italics mine).

It is obvious that Sagan, while knowing nothing about Christian Science, is willing to be fair. The italicized sentences would be accepted by any health-providing practitioner, whether physician or Christian Scientist. Neither the medical arts nor Christian Science as practiced by humans has a perfect record of saving lives. Martin Gardner, while knowing nothing about Christian Science, wrote a totally misleading, deceptive description of what Sagan had said.

Nowhere does Sagan mention insulin, diabetes, or antibiotics in regard to Christian Science practice. Indeed, the only reference Sagan makes regarding Christian Science is the one quoted. Nowhere else in the index is the subject found. Nor is insulin, diabetes or antibiotics. Where, then, does Gardner get the authority for what he said on these subjects? It is my firm conviction that Gardner used the opportunity to inject a personal criticism he holds for Christian Science. That is unprofessional and a discredit to the *Post* for allowing it. I would inform Gardner that Christian Scientists love, protect, and provide for their children just as much as, possibly more than, does he for his. Their use of Christian Science treatment for themselves and for their children, instead of medicine, is based on their experience with its efficacy, something unknown to Gardner.

Is it the *Post's* policy to print any review without regard to the accuracy of the critique as related to the book reviewed?

W. L. Williams
Manassas

My reply:

W. L. Williams should read Sagan's book rather than rely on its index. On page 9 Sagan writes: "There is still a religion, Christian Science, that denies the germ theory of disease; if prayer fails, the faithful would rather see their children die than give them antibiotics." In 1989 eleven-year-old Ian Lundman died after his Christian Science mother refused to give him insulin for his diabetes. Last January the Supreme Court upheld a $1.5 million judgment won by the child's father. There have been dozens of similar cases in the last few decades.

As for my "knowing nothing about Christian Science," William should read my book *The Healing Revelations of Mary Baker Eddy*.

In regard to Robert Ashford's letter: Homeopathy, for true believers, is a dogma held as firmly as the faith of a Christian Scientist. I wonder if Ashord knows that Mary Baker Eddy, before she discovered Christian Science, was an enthusiastic user of homeopathic drugs. She later decided it was faith in the drugs (the placebo effect), not the drugs, that cured. Anyone interested in my critique of homeopathy can consult the chapter on medical cults in my paperback *Fads and Fallacies* and my chapter "Water with a Memory?" in my 1992 book *On the Wild Side*.

PAUL EDWARDS'S
REINCARNATION

"Religious insanity is very common in the United States."

—Alexis de Tocqueville, as quoted by Paul Edwards

Because reincarnation is a fundamental doctrine of Hindu religions and most forms of Buddhism, there are probably more people around the world who believe it than those who prefer the Judaic-Christian-Muslim view that our lives begin at birth and will continue after death in heaven or hell. In recent decades New Age infatuations with Eastern faiths have impelled vast numbers of Americans to see in reincarnation the satisfaction of their hunger for immortality. A 1991 Gallup poll found that 1 in 5 Americans are reincarnationists. More surprising, 14 percent of Catholics and 19 percent of Protestants expressed belief in reincarnation. This is truly astonishing because nowhere

This review first appeared in *Free Inquiry* (summer 1997).

does the Bible defend the doctrine. Indeed, reincarnation sharply contradicts Christian teachings.

A raft of books defending reincarnation has been written by Americans and published by firms whose owners and editors have not the slightest interest in the doctrine beyond its power to inflate profits. Now, for the first time in history, in any language, a distinguished philosopher—in addition to his books and papers, Paul Edwards edited the eight-volume *Encyclopedia of Philosophy*—has written a comprehensive, powerful attack on reincarnation.

But *Reincarnation: A Critical Examination* (Prometheus, 1996) is much more than an attack. Its prose is vigorous and entertaining, its arguments cogent, its scope awesome, and its documentation impeccable. Edwards seems to have read everything on the topic, to have considered every argument pro and con.

True believers in reincarnation probably will not read this book, and among those who do, not many are likely to alter their opinions. It is a rare event when believers of any stripe change their minds about anything. On the other hand, perhaps a few with brains still open to reason will find Edwards's rhetoric persuasive. For anyone, his book will be a delight to read and a basic reference to own about one of the strangest religious phenomena to befuddle America since Christian Science.

Many pages are devoted to the claim that reincarnation provides, as Edwards himself believes, a better solution to the terrible problem of irrational evil than the traditional Christian one. Catholic and Protestant fundamentalists still believe that all of us are sinners as a result of Adam and Eve disobeying a command of Jehovah. They also must believe that God was so disappointed with his own creation that he drowned all men, women, babies, dogs, and cats except for Noah and his unremarkable family. They must agree with St. Paul that the only way to escape eternal damnation is to believe that Jesus came to Earth to release humanity from God's awful curse. Moreover, Paul added, to be born again and saved from hell one must also believe God raised Jesus from the dead. Can you imagine a deity more distant from the loving God taught by the historical Jesus?

Reincarnation does away with such a cruel creator. The horrors of hell are replaced by karma, a law as much a part of nature as the laws of gravity. Indeed, many reincarnation-ists were and are atheists, such as the British philosopher John McTaggart Ellis McTaggart, who sees no need for a god to uphold karma. We existed before in different bodies. We will continue after death in other bodies. Every evil deed will be punished. Every good deed will be rewarded. Irrational suffering no longer demands justification by reasons we cannot now comprehend. Karma ensures that justice will always prevail, if not in this life, in lives to come.

Who has not longed to participate in future events on Earth, to take part in building a better world, to see with human eyes what happens to one's descendants, to observe the changes science and technology will bring? Christians, Jews, and Muslims are unable to participate in Earth's future history. The best they can do is observe it from a distant heaven. Reincarnationists have the hope of actively participating in the planet's future.

Foremost of all objections to reincarnation is the diffi-culty of conceiving how we can be the same person who lived before if we have no memory of an earlier life. Edwards quotes Leibniz: "What good would it do you, sir, to become King of China on condition of forgetting what you have been? Would it not be the same thing as if God at the same time he destroyed you created the king of China?" How can we profit from past experiences if we can't remember them?

Here and there, especially in India, a few rare individ-uals claim to recall portions of a past life, such as the famous case of the American Bridey Murphy. Edwards thoroughly demolishes these flimsy claims. They are usu-ally based on hypnotic sessions during which a hypnotist consciously or unconsciously guides the sleeping patient with leading questions. Not only is hypnotism used today to "regress" patients to former lives, but some reincarnation therapists, notably the former Baltimore dentist Dr. Bruce Goldberg, "progress" patients to observing events in their future incarnations. Most reincarnation therapists treat patients for causes in past lives. Dr. Goldberg treats patients for causes they will experience in future lives!

Many reincarnationists claim that everyone will remember previous lives during a period between incarnations that Edwards calls the "interregnum," a time when our soul or "astral body" exists in a physically bodiless state. Believers differ on how long interregnums can be. They may be as short as a few days or as long as hundreds of years. Perhaps after many incarnations, or at the end of them, if there is one, our brains will recall everything.

One of the strongest objections to reincarnation is that, in India, where almost everyone believes in karma, efforts to alleviate suffering are dampened by the belief that the Untouchables are being punished for past sins. As Edwards makes clear, if you truly think that, say, a child dying of cancer is suffering from sins in a former life, why should you try to thwart karma by easing the child's pain? Edwards quotes Christmas Humphreys, an attorney who founded the British Buddhist Society: "He who suffers suffers from his deliberate use of his own free will." We should not show sympathy for "cripples, dwarfs, and those born deaf and blind," Humphreys adds, because their afflictions "are the products of their own past actions."

Reincarnationists are troubled by this reasoning. Some insist that you should do your best to relieve pain because if you succeed it proves a person's punishment is not as severe as it seemed! But this goes against the notion that karma is an inexorable law that cannot be altered. If we can change it at will it turns karma into what Edwards calls a "vacuous" law. In any case, the doctrine of karma continues to be a major obstacle to efforts by India's government to make the miserable less miserable.

Another objection to reincarnation explored by Edwards is what is known as the "population problem." If everyone alive today once inhabited a previous human body, how can the world's population explosion be explained? There simply are not enough former earthlings to account for today's population, and each year the problem gets worse. Edwards considers what he calls several "noxious ad hoc assumptions" put forth to explain this. One is that we had past lives as animals, as expressed in Langdon Smith's popular ballad "Evolution" which begins· "When you were a tadpole and I was a fish, in the Paleozoic time." Another is that we lived before on other planets, or in a densely populated "astral world" on some higher plane of reality. The most bizarre conjecture comes from Dr. Goldberg. He solves the population problem by assuming that a soul can occupy two or more bodies at the same time!

Still another powerful argument against reincarnation rests on the sudden deaths of tens of thousands of persons in an earthquake. A reincarnationist must believe that all the victims were simultaneously punished for past sins. "How did this nonintelligent principle [karma] set up the geological forces so as to achieve the desired result with complete precision?" Edwards asks.

If the law of karma is infallible, Edwards wants to know, how can a reincarnationist bring himself to believe that the six million Jews exterminated by the Nazis all deserved their fate? The same goes for Inquisition victims, and those of Stalin. "Since the Jews deserved extinction," writes Edwards, "the Nazis were not really criminals. . . . I assume that Eichmann deserved to be hanged, since he *was* hanged, but the many Nazis who escaped deserved to escape." The most powerful objection to the death penalty becomes meaningless because, if karma holds, no innocent man can be executed. "People may indeed be innocent of the crime with which they are charged, but if they are executed this is what they deserved. It makes one dizzy."

Suppose a child is run over by a car and killed. What

should a reincarnationist pastor say to console the mother? Edwards imagines him saying: "It all makes sense—your child deserved her fate; she sinned in a previous life, and in view of the severity of her suffering we may assume that her sins were enormous. What is more, you yourself are acutely suffering and there is no doubt that you are being punished for some serious transgression either in this or an earlier life or both."

"If I were the mother," Edwards continues, "and a baseball bat were handy, I would hit the karmic pastor over the head and, as he screams with pain, I would say: 'You deserve your pain not because of a sin in your previous life but because you are a monster right now.' "

One reason Edwards's book is such a pleasant read is that he has a sense of humor and sarcasm worthy of Voltaire or H. L. Mencken. I have cited some instances. Here is one more. Edwards is considering the view of many reincarnationists that evil individuals can incarnate not only as animals, but also as insects, plants, or even rocks and jars. Tongue in cheek, Edwards speculates on the possible former animal lives of some prominent persons:

> It is widely believed that the poet Edith Sitwell was a flamingo in an earlier life and there cannot be a serious doubt that Winston Churchill had once been a bulldog. Bull terriers, the lovable little dogs whose noses look as though they had been bashed in, were probably prize fighters in a previous life. As for Marlene Dietrich, the general consensus now is that she once was an emu. There seems to be no other way of explaining her treatment of her daughter, Maria Riva. J. Edgar Hoover was almost certainly a praying mantis and the same is probably true of Richard Nixon and his criminal associates who brought us Watergate.

The book's funniest chapter is about astral bodies in which Edwards covers in hilarious detail two famous cases of "astral projection." A century ago S. R. Wilmot, on a stormy crossing of the North Atlantic, had a vivid dream that his wife's astral body had visited his stateroom. Mrs. Wilmot, then at home in the United States, appeared by her husband's bunk, wearing a nightgown. As he told it, "she stooped down and kissed me, and after gently caressing me, quietly withdrew." When he returned home, his wife staggered him by telling about her out-of-body trip. She even recalled unusual features of the stateroom, and the presence of a passenger in the upper berth.

The other case involves Dr. George Ritchie, a Virginia psychiatrist. In 1943, when he was pronounced dead in an Army hospital, his astral body floated into heaven where he encountered Jesus. The Lord then took him on a tour of both hell and heaven. Ritchie describes the tour in dazzling detail in his 1988 book *Return from Tomorrow*. Edwards notes that, the closer one gets to the original documents of such notable cases, the greater become the discrepancies. This "hiatus between the original claim and the actual evidence," Edwards calls the "Ritchie-Wilmot syndrome."

Astral body defenders all agree that each of us has within us an exact replica of our physical body, a double who occasionally roams about at will. Curiously, when astral bodies are seen, they are invariably clothed. Many astral experts, Annie Besant for example, contend that every physical object has its counterpart in the astral world. This requires, writes Edwards,

> that every time somebody produces something, he also produces an astral copy of that thing. A carpenter who builds a set of bookshelves is really building two sets, the regular one he sells to his customer and an astral copy he sells to nobody. And the same of course applies to everything. A dentist, for example, who fills a tooth is really filling the tooth he thinks he is filling as well as its astral duplicate, and when I am writing these lines I am really writing them twice at the same time. This is too much. I would rather believe that all astral bodies are always naked and that we are deluding ourselves when we observe them

clothed. If it were not needed for reincarnation one might almost be tempted to give up the astral body.

It would require too much space to discuss Edwards's astute comments on the views of the two most noted philosophers who defended reincarnation—McTaggart, mentioned earlier, and the French-born American Curt John Ducasse. Among lesser believers trounced in the book are Elizabeth Kübler-Ross ("the most credulous person who ever lived"), Stanislaw Grof (the second-most credulous), Annie Besant, Edgar Cayce, Henry Ford, General George Patton, Princess Diana, and such entertainers as Shirley MacLaine, Loretta Lynn, and Sylvester Stallone.* "Stallone thinks he may have been a monkey in Guatemala," Edwards reports, "something I find entirely credible."

Ian Stevenson, a psychiatrist and parapsychologist at the University of Virginia, is far and away the most famous living reincarnationist and the person most tireless in seeking empirical support for the doctrine. Stevenson believes almost everything on the psi scene. He thinks Chicagoan Ted Serios could project his thoughts onto Polaroid film. He believes that Pavel Stepanek, a resident of Prague, was a great clairvoyant until he lost his powers. He has written serious papers about how the dreams of dozens of people foretold the sinking of the Titanic.

Edwards's attacks on Stevenson, whom he believes to be "sincere but deluded," are impersonal but harsh and unrelenting. It would be fascinating to read Stevenson's review of Edwards's book, assuming he has the courage and chutzpah to review it.

---

*And Norman Mailer. According to *Time* (September 30, 1991, page 69), he told an interviewer: "I happen to believe in it [reincarnation]. . . . It just seems to me that if we lead our lives with all that goes wrong with them, and then we die and that's the end of us, that doesn't make much sense."

BEHIND THE
CRYSTAL BALL

Recent decades have seen an astonishing upsurge of popular infatuation with fringe science and the paranormal. A hundred years ago, no newspaper in the United States carried a horoscope. Today almost every paper except the *New York Times* has such a feature. A recent U.S. president and his wife, Ronald and Nancy Reagan, were firm believers in the influence of stars on human events. Polls show that about half of all Americans believe in Satan, angels, demons, extrasensory perception, precognition, and alien spacecraft invading our skies.

How does this compare with people's beliefs in earlier ages? It is the opinion of Anthony Aveni, in his entertaining history of Western occultism (*Behind the Crystal Ball*, Random House, 1996) that until the rise of modern science a belief in magic was rational, mainstream common sense. Not until after Galileo and Newton was

---

This review first appeared in *Nature*, January 30, 1997.

such magic demoted to the fakery of stage conjuring. This shift surely has occurred among those knowledgeable about science. But has it also taken place among the masses whose knowledge of science could be scribbled on a postcard?

Two-thirds of Aveni's book can be seen as a skilful summary of Lynn Thorndike's multivolumed *History of Magic and Experimental Science*, and one-third as taking up where Thorndike left off. The volume is amusingly illustrated, its bibliography extensive, and its facts carefully documented by footnotes. No recent popular history of occultism has covered such a wide swath.

Aveni begins with the "magic" of ancient Egypt and Mesopotamia, followed by Greece and Rome. After a colorful romp through the Middle Ages and Renaissance, he turns to the scientific rubbish of the nineteenth century. There are excellent accounts of the origin of modern spiritualism in the toe rappings of the Fox sisters, the levitations and other paranormal feats of the medium D. D. Home, Madame H. P. Blavatsky's theosophical twaddle, and the worldwide spread of phrenology.

Chapters on twentieth-century "magic" include the emergence of parapsychology, the spoon-bending of Uri Geller, crystal power, UFO abductions, dowsing, channelling, near-death experiences, alternative medicines, and many other marvels.

Although Aveni, a professor of astronomy and anthropology at Colgate University in Hamilton, New York, clearly does not believe any of the humbug he writes about, for some reason he feels obliged to be objective, to give balanced accounts of everything. "My goal will not be to attack alternative ways of perceiving reality," he writes. "I have no desire to pronounce all magic superstitious flotsam—science gone awry. Nor do I wish to demystify magic and reduce it to a set of explanations that are inferior to my own scientifically trained way of understanding the world."

I find it incredible that an astronomer would not wish

to say that belief in the magic he covers so well in his book is inferior to his beliefs as a scientist. It is this tinge of cultural relativism—springing no doubt from Aveni's anthropologist's hat—that, to an old debunker like me, detracts from the value of his history. For example, although Aveni surely knows that Ted Serios, the Chicago bellhop who projected his thoughts on to Polaroid film, is not a genuine psychic, he cannot bring himself to say this outright. Instead, after mentioning that magician James Randi can duplicate all the feats of Geller and Serios, he adds: "Does this prove that Geller and Serios are fakes? In other words, does the possibility of fraud mean that it actually had occurred?" Later, Aveni soberly considers the theory of some parapsychologists that spoon-bending can result from the mind using quantum laws to alter metal.

Aveni certainly agrees with Houdini that Home was "a hypocrite of the deepest dye" yet, after publishing Robert Browning's damaging account of a Home seance, he has to "balance" it with Mrs. Browning's high praise of Home. One longs for Aveni to abandon his curious notion that science and magic are somehow equally valid ways of seeing the world.

I would have liked to see some mention of Chinese and Hindu astrology, each of which is based on star patterns entirely unlike those of Western astrology. (If one is right, the other two are wrong.) Aveni's discussion of alchemy misses the intense labours of Isaac Newton. A chapter on anthropology deals with the sorcery of primitive cultures, but is silent on Margaret Mead's naive beliefs in magic and UFOs, and on modern anthropologists and sociologists who are deep into the paranormal. There is nothing about St. Augustine's well-reasoned attacks on astrology, or the bashing of psychic charlatans by the Greek satirist Lucian.

I think the gulf between the opinions of intellectuals and the superstitions of the populace was not much wider in past ages than it is today. Is a person who takes homeopathic remedies, knowing they contain not a single molecule of the original substance, much different from a medi-aeval peasant who tried to cure an ailment by drinking the blood of a frog? Is there much difference between those in past centuries who had visions of the Virgin Mary, angels and devils, and today's devout Catholics who have similar visions? Every time there is a report of a Mary statue that weeps or bleeds, the church is invaded by hordes of the faithful eager to witness the miracle.

Tens of thousands of Americans now regularly telephone "psychics" for information. Do they differ from ancient Romans who sought similar advice from the patterns of animal livers and entrails, or from the voices of oracles?

For evidence that the gap between superstition and science is almost as wide today as it ever was, compare the number of books on the "New Age" shelves of bookshops with the number of books about science. Only the kinds of superstitions have altered. And in some cases, such as astrology, not even that has changed.

THE STORY OF A CHESS PRODIGY

Y ou don't have to play chess to relish Fred Waitzkin's
  Searching for Bobby Fischer (Random House, 1988),
but if you do know the game, you'll find it even harder to
put down. No other book has so colorfully captured the
psychological, social, and political aspects of the Royal
Game: the wealthy Russian grandmasters, supported by
the state and adored by the masses; U.S. grandmasters,
eking out a living with jobs they hate; chess bums who
hustle "fish" (mediocre players) in public parks and seedy
chess clubs; and the lives and personalities of amazing
children who get hooked by the game.

Running through the book is the question that troubles
all parents of chess prodigies. Were they wise to teach
their son chess? Will he become an honored grandmaster,
as happy and well adjusted as the Russian Boris Spassky,
or will the game turn him into a miserable misfit like
Bobby Fischer?

This review first appeared in the *News and Observer* (Raleigh, N.C.),
December 4, 1988.

Fred—as his son Joshua, since the age of three has called his father—taught Josh the game when he was six. The handsome little boy, with big brown eyes, loved the game. In no time he was spending his free-from-school time in Washington Square Park, in Greenwich Village, playing blitz chess (games played with extreme rapidity) with the park's hustlers. Soon he was trouncing his dad, a New York freelance writer, and rising quickly in tournament play with other children. Now eleven, he is the nation's top-ranking player for his age.

Bobby Fischer, another child prodigy who once battled the Washington Square chess bums, haunts every chapter. It was his much publicized defeat of Spassky in Iceland that inspired thousands of young men (some women, too) to take up the game, including Fred. What has happened to the great Fischer? The book's saddest chapter is devoted to this tragedy.

After beating Spassky, Fischer stopped playing tournament chess to vanish into the slums of Los Angeles. He is one of the most striking examples in recent times of an intellectual giant who, given a different personality and upbringing, could have become an eminent mathematician or theoretical physicist. Instead, his vast intelligence became focused down on those sixty-four little squares where wooden pieces get pushed about in obedience to rules that have no bearing on anything else in life. In the subculture of chess, Bobby is a genius. Outside that subculture, he is an imbecile.

I once thought Fischer's refusal to play tournament chess on Saturday was because he is half-Jewish, but no. It was because he became a convert to the late Herbert Armstrong's World Wide Church of God, a moronic cult that worships on Saturdays and expects the Second Coming any day now. Fischer gave the church thousands of dollars until he had a falling out with Herbert's son Garner Ted. Garner was the cult's heir apparent until the old man excommunicated him for such widespread womanizing that it makes Jim Bakker and Jimmy Swaggart seem like models of monogamy.

Fischer's current obsession is hatred of Jews. He has developed a burning admiration of Hitler and the Nazis. The holocaust? It is a fantasy concocted by evil Zionists. He won't read *Chess Life*, the leading chess magazine, because he thinks Jews run it the way they run the U.S. Chess Federation and the U.S. government. His paranoia is growing. The KGB is after him because he might take the Soviet chess leadership away again. He won't watch TV — it emits deadly rays. Recently he had all his fillings removed, leaving holes in his teeth. He is afraid the fillings will pick up radio transmissions and dangerous vibrations.

Fred tried to find Bobby in L.A. but never did. Nor have private detectives hired by magazines, or persons seeking to drag him back into tournament play. He has become a recluse, a shabby street bum reportedly playing better chess than ever, but unwilling to compete. Chess players of course still idolize him. A New York woman player told Fred that she speaks to Fischer regularly on the phone. "He's so pure, like Jesus," she said, oblivious of the irony of likening Bobby to a Jew.

In 1986 the Russian champion Gary Kasparov visited the Bronx to play simultaneously against fifty-nine youngsters. Josh was among the two who achieved draws. Afterward he told a TV interviewer that when he grew up he wanted to play second base for the Mets. Fred is not sure whether that was good or bad news. We shall have to wait and see.

POSTSCRIPT

In 1993 Waitzkin's book was made into a memorable movie with the same title as the book. I have not kept up with Josh's career since then, except to learn that when he was nineteen he was awarded the Samford Chess Fellowship, which gave him $32,000 a year to study and play chess.

JIM BAKKER AND TAMMY FAYE

J im Bakker's massive confessional *I Was Wrong* (Thomas Nelson, 1996) is a remarkable work in many respects, not least of which is the skill with which it is written. His earlier autobiography, *Move That Mountain* (1976), was a ghostwritten potboiler. *I Was Wrong* acknowledges the help of Ken Abraham, "who used his skills to craft this book." (The last of his many thanks is "to my best friend who never left me . . . Jesus!") However it was crafted, the book is Bakker's solemn effort to tell the story of his sudden rise to be one of the nation's most successful televangelists, and the sudden downfall in which he lost his ministry, his freedom, his self-respect, and his wife.

For readers unfamiliar with the sad soap opera of Jim and Tammy Faye, here is a quick rundown. The diminutive pair—she is four feet eleven, he five feet eight—met at North Central Bible School, in Minneapolis. Both came

---

This review first appeared in the *New York Review of Books*, May 29, 1997.

from strict Pentecostal homes in which life centered on church services. Pentecostals are evangelicals who take their name from the biblical account of the seventh Sunday after Easter, when the Holy Ghost descended on the Apostles. They believe that the gifts of Pentecost, such as instant faith healing and the power to speak in tongues, remain in effect today. In her 1978 autobiography, *I Gotta Be Me*, Tammy tells how she was born again at age ten when she responded to an altar call. "I lay on the floor for hours and spoke in an unknown language. . . . I was walking with Jesus."

Jim proposed on their third date. Tammy, then eighteen, giggled throughout the ceremony. Because the Bible School banned student marriages the young newlyweds left to hit the road as traveling soul savers. Jim preached. Tammy played the piano and accordion and sang. In an effort to reach children with the gospel, Jim and Tammy began using puppets. Tammy had a tiny voice for animating Susie Moppet and a deep voice for Allie Alligator. Sunday school meetings tripled in attendance. Word of their popularity reached Pat Robertson, who invited them to do a children's show on his fledgling Christian television station in Portsmouth, Virginia. *The Jim and Tammy Show* was a huge success.

In *I Gotta Be Me* Tammy says Jim worked so hard on their show that he had a mental breakdown. For a month he stayed in bed, Bible in hand, muttering, "Please God, don't let me lose my mind." Once recovered, Jim persuaded Robertson to let him start a show modeled on Johnny Carson's. It became CBN's *700 Club*. Jim was host of the show for years before Pat began taking over every other night. Rifts between the two widened. "I loved him [Pat] with a real deep love of the Lord," Tammy writes. "But at times he would do certain things. . . . I built up a terrible, terrible resentment in my heart against him."

Pentecostals frequently hear, sometimes even see, the Lord speak to them in mysterious ways. Jim and Tammy liked to open the Bible at random and read the first verse they saw. Several times the Bible opened on Ezekiel 12:1–6 ("Prepare thy staff for removing"), which they interpreted to mean that God was telling them to leave CBN. For a brief time in California they were associated with televangelists Jan and Paul Crouch. It was there they started the PTL Club, which people could join by sending money. The letters stand for Praise the Lord, but detractors were soon calling it Pass the Loot. Quarreling broke out, especially between Tammy and Jan, and Jim and Tammy were back on the road.

The pair resumed broadcasting on behalf of the PTL Club from a small studio in Charlotte, North Carolina. Featuring famous actors and entertainers as guests, it quickly became the most widely watched Christian show on television. Jim proved to be a brilliant performer, not only at chatting with guests and creating a friendly mood but also at increasing the steady flow of donations. He would weep about PTL's needs, and thousands of elderly ladies (it was said) would turn to cat food so they could send their dollars to the Lord. Testimonials of instant healings poured in as Jim would get what Pentecostals call the "word of knowledge"—insights from the Holy Spirit that enabled him to describe specific ailments of listeners.

As PTL's many bank accounts grew by the millions, Jim's dreams grew more grandiose. He hired Roe Messner, a devout Pentecostal church builder, to construct Heritage USA, a huge Christian theme park in Fort Mill, South Carolina. By 1986 PTL was reaching over twelve million households by way of its satellite network. That was the year that bad publicity blasted Jim's empire.

The *Charlotte Observer*, long an enemy of PTL, disclosed that seven years earlier Jim had committed adultery with a sultry gospel groupie named Jessica Hahn, then the secretary of a Pentecostal church in New York. She claimed she had been a virgin until Jim raped her. To keep her quiet, Bakker's co-pastor Richard Dortch paid her

$250,000. Later payments of hush money, all from PTL donations, brought the total to $363,700. Televangelists were outraged. One of the more prominent, John Ankerberg, joined the attack by accusing Jim of frequent homosexual episodes. Jimmy Swaggart called PTL "a cancer on the body of Christ."

The U.S. Department of Justice began looking more seriously into how PTL handled its funds. A grand jury indicted Bakker on twenty-four counts of fraud and conspiracy. He was accused of bilking his followers out of $158 million, from which $3.7 million went to maintain his and Tammy's opulent style of life. The Bakkers owned several expensive vacation houses. They had a fleet of high-priced cars. Half a million went to buy and decorate a condo in Florida. Childish extravagances came to light. Tammy's closet was 25 by 25 feet. A Christmas tree cost $5,000. Jim once sent a security guard to buy $100 worth of cinnamon buns just so he and Tammy could smell them in their hotel suite. Not a bun was eaten. When clothing and other personal items were moved from Bakker's parsonage in Tega City, South Carolina (the house burned down in 1990), the move was made in a private jet that cost PTL $105,000.

Jim asked Jerry Falwell, the Baptist fundamentalist who ran *The Old Time Gospel Hour* from his church in Lynchburg, Virginia, to take temporary custody of PTL. It was a bizarre request because Falwell has no use for Pentecostal doctrines and once described speaking in tongues as similar to the stomach rumblings of someone who ate too much pizza. His Moral Majority was much more concerned than the Bakkers to support the right wing of the Republican party on such matters as abortion and pornography. By comparison, the Bakkers are largely nonpolitical religious entertainers. At first Jim trusted Falwell, but he soon became convinced that Falwell's secret motive was to have PTL declared bankrupt so he could purchase its valuable network. The bankruptcy occurred, but not the takeover.

Heritage USA and Bakker's Inspirational Satellite Network are now owned by California evangelist Morris Cerullo.

Jim's trial lasted almost three years. A panic attack in court, during which he curled up in his chair, sent him for a time to a mental hospital. After Bakker pleaded innocent, the jury found him guilty on all twenty-four counts. Judge Robert Potter, known as "Maximum Bob" for his harsh sentences, gave Jim forty-five years in prison and fined him half a million dollars.

Reverend Dortch, after a plea bargain of guilty, was given eight years and fined $200,000. "I could not believe that I participated . . . in doing what I knew was wrong," he said, as he wept in court and begged for mercy. "I failed my Master, failed my family, and failed myself." He served only sixteen months. Since his release he has written three mea culpa books: *Integrity: How I Lost It* (1990); *Losing It All and Finding Yourself* (1993); and *Fatal Conceit* (1993).

Jim ended up in a federal prison in Rochester, Minnesota. Alan Dershowitz, hired by Roe Messner, managed to get Jim's sentence reduced to eighteen years. It was later reduced to eight. After five years in prison Jim was out on parole. He settled in Hendersonville, North Carolina, where his lawyer Jim Toms lives. It was there in a rented farmhouse that he wrote *I Was Wrong*.

Tammy stood by her man until early in 1992 when she told him she planned to dump him and marry Roe Messner. Jim says he was devastated. In his book he describes Roe as a devout Pentecostal who became his "best friend"—aside from Jesus. He writes that he felt stabbed in the back. Reverend Dortch told *People* magazine in April 1992 that Roe "should hide his face in shame and put on sackcloth and ashes. . . . I have Christian contempt for Roe Messner, who took advantage of his so-called best friend's wife."

I Was Wrong is Jim's grim account of his downfall, his life in prison, his frequent depressions, his feeling of being abandoned by both God and Tammy, and his final rebirth when he persuaded himself that imprisonment had been

God's plan all along to make him a better man. "I was wrong" occurs like a refrain throughout the volume, very much in the evangelical tradition of public penitence which can lead to public redemption. Bakker admits to many mistakes and sins, though not to any of the crimes for which he was sentenced. His greatest "wrong," he says, was his excessively expensive and showy style of life.

American televangelists have long cherished what has been called the "prosperity gospel." Oral Roberts likes to tell how he opened a Bible one day and his eyes fell on John 3:2: "Beloved, I wish all things that thou mayest prosper and be in health, even as thy soul prospereth." The verse had a tremendous effect on Oral. God does not want any of His children to be poor! "Seed faith" was Oral's term for money given to his ministry. The more generous you are, he promised, the more money God will return.

It was not only Pentecostal evangelists who preached a "health and wealth" gospel, but also liberal "feel-good" ministers such as Norman Vincent Peale, Robert Schuller, and Harlem's Reverend Ike. It was easy to rationalize their own conspicuous prosperity. Did it not prove to their flock how well seed faith works?

In one of his later chapters, Bakker reveals, with no hint of humor, that when he reread the gospels in prison he was dumbfounded to discover that Jesus hated the rich. "To my surprise, after months of studying Jesus, I concluded that He did not have one good thing to say about money." Bakker ticks off the relevant passages: "Lay not up for yourselves treasures upon earth . . ." (Matthew 6:19). "But woe unto you that are rich . . ." (Luke 6:24). Jesus advises a wealthy young man to sell all his possessions and give the money to the poor. He said it was harder for a rich man to enter heaven than for a camel to go through the eye of a needle.

Bakker writes that for years he taught that there was a low arch in Jerusalem called the "camel's eye." Camels had to kneel to pass under it, suggesting it was difficult but not impossible for the rich to be saved. Jim consulted references and found "not a shred" of reputable evidence for this conjecture. What about John 3:2? Bakker checked the original Greek. He was amazed to learn that the word translated as "prosper" had no reference to wealth. "I had to face the awful truth that I had been preaching false doctrine for years and hadn't even known it!" He now feels he was wrong to flaunt his wealth. He no longer accepts Tammy's frequent justification: "We were worth it."

From the beginning Jim has viewed his brief fling with Jessica as a sin. It occurred, he writes, at a time when he and Tammy were at odds and she had turned to Gary Paxton, a long-haired country singer with a bushy dark beard who taught Tammy how to improve her voice, and who wrote gospel songs for her. Both Paxton and Tammy deny that sex was involved, but Bakker thought otherwise. He claims his affair with Jessica was intended to make Tammy jealous. Jessica, he writes, took the initiative. He was so ashamed that he never had the courage to tell Tammy about it, or about the hush money Dortch paid to keep Jessica from blabbing.

Tammy's romance with Paxton was not the only time she strayed. On a later occasion she found another lover, PTL's musical director Thurlow Spurr. Jim learned about it when he discovered in the garbage a love letter Tammy had written, then torn into tiny bits. It took Jim hours, he writes, to assemble the pieces. Like Paxton, Spurr was fired.

Bakker's account of his years in prison tells of close friendships with inmates, of his loneliness, of his bouts of depression. His job was cleaning toilets, a task he didn't mind because he knew they would be clean when he used them. He talked daily with Tammy by phone. He lifted weights in the gym. He took up painting in watercolors. Visits by Billy Graham and his son Franklin boosted his sagging selfesteem. He fended off attempted rape by an unnamed burly inmate. The man was eventually transferred to another prison, though not because of Bakker, who was faithful to prison ethics, and would not "snitch."

155

For the first time Bakker reveals that as a child he had been repeatedly abused by a friendly young man without realizing that what he was doing was sexual. He wonders: Can I be a repressed homosexual? The prison doctor assures him he is not. In a candid passage he tells how inmates tried to teach him how not to walk like a girl or toss a ball like a girl. He had never handled a baseball as a boy. For a time his cellmate was Lyndon LaRouche Jr., the eccentric right-wing political figure who was doing time for tax evasion and for swindling contributors of $30 million.

When I listened to George Bush give a televised convocation speech at Jerry Falwell's Liberty College, and refer to Falwell as a dear friend, I assumed he was doing his best to curry favor with the religious right. After reading Bakker's book I am now convinced that Bush wasn't faking. Here is how Bakker reports a private conversation with Bush in 1986:

> Consequently, I was not surprised when, shortly after our lunch began, the vice president brought up the subject of Christianity. I quickly realized that the future presidential candidate was interested in how he might make inroads with the evangelical Christian voters. In the presence of only his personal aide and my assistant, George Bush probed, "Jim, I believe in God, but I am just not comfortable with the term 'born again.' That is not a term we use in our church tradition."
>
> "Well, do you believe the Bible is the inspired Word of God?" I asked.
>
> "Yes, I do."
>
> "Do you believe that Jesus Christ is the Son of God, that He was born of a virgin, and that He died on the cross for our sins? Do you accept Him as your Lord and Savior?" I asked the vice president.
>
> "Yes, yes, I do," George Bush replied.
>
> "Well, that's what being born again is all about."
>
> The vice president and I went on to talk about prayer, and he emphatically declared that a day never went by when he did not ask for God's help and wisdom in dealing with his responsibilities.

There are glaring omissions in *I Was Wrong*. Bakker never mentions the lies he told his television audience. He said there was a strict limit on the number of PTL lifetime partnerships—for a thousand dollars a partner was guaranteed three nights free every year at Heritage Grand Hotel—when actually he was selling partnerships far beyond the luxury hotel's capacity. Nor does Bakker mention the time he said tearfully that he and Tammy had given "every penny" of their life savings to PTL. A few days later he made a $6,000 down payment on a houseboat. Nor does he recall saying that copies of a statue of David and Goliath that he was offering to donors "might well be worth $1,000." They cost PTL ten dollars each.

Bakker has nothing to say about Charles Shepard's carefully documented 635-page history, *Forgiven: The Rise and Fall of the PTL Ministry* (1989). Shepard was the *Charlotte Observer*'s reporter who for years covered the PTL scandals and showed how much money Bakker was making. His articles won him a Pulitzer Prize in 1988. Nor is there a trace in Bakker's book of Art Harris's interview (*Penthouse*, January 1989) with John Wesley Fletcher, a flamboyant Pentecostal Bible-thumper and faith healer who often appeared with Jim on PTL shows. An alcoholic, frequently in trouble with the law, he was the man who introduced Jim to Jessica.

"I was Jim Bakker's male prostitute," Fletcher told Harris. He described Jim as a bisexual who lusted after young male cameramen. On three occasions, Fletcher reported, he had had oral and anal sex with Bakker.*

---

*The actor Efrem Zimbalist Jr. attributed his religious conversion to Fletcher's persuasive urgings on *The Jan and Paul Crouch Show*. Zimbalist later became a member of PTL's board of directors. Ex-Pentecostal minister Austin Miles, in his revealing book *Don't Call Me Brother: A Ringmaster's Escape from the Pentecostal Church* (Prometheus, 1989), recalls asking Zimbalist if he felt embarrassed when he learned of Fletcher's homosexual relations with Bakker. "Not at all," said the actor. "No matter what he turned out to be, I'll always be grateful to him. It's because of him that I'm a born-again Christian and know Christ."

Fletcher tells of catching Bakker in bed with David Taggart in a Bermuda hotel room. As Bakker's top assistant, Taggart received an annual salary of over $400,000, not counting bonuses. He and his brother James, PTL's interior decorator, were convicted in 1989 of evading more than half a million dollars in taxes. Each was sentenced to seventeen years in prison and fined half a million dollars. Jim is silent about their arrests.

After Fletcher and others told their stories at a secret meeting of leaders of the Assembly of God, the large denomination to which Bakker and Dortch belonged, Bakker was defrocked for adultery and alleged bisexual activities. Jimmy Swaggart was later excommunicated by the same denomination for hankering after prostitutes.

Jessica lied, Fletcher told Harris, when she claimed to be a virgin raped by Bakker. He himself, he said, had slept with her earlier as well as immediately after she had slept with Jim. Jessica, by the way, was befriended by Hugh Hefner, owner of *Playboy*, who paid for plastic surgery on her breasts and face. She was featured nude in two issues of *Playboy*, one showing her before the surgical enhancements (November 1987), the other showing her after (September 1988).

Can Fletcher's charges in *Penthouse* be trusted? We will probably never know for sure. Tammy branded them a "sick lie." In 1988 Fletcher tried unsuccessfully to hang himself with a lamp cord. In 1990 he was sentenced to three years probation for perjury in his testimony to a grand jury about Bakker's liaison with Jessica Hahn. On June 2, 1996, The *Globe* tabloid ran a pitiful picture of Fletcher in a hospital bed and said he was dying of AIDS. The paper added, without producing any evidence, that he knew he was HIV positive when he had sex with both Jessica and Bakker.

Fletcher was not the only PTL employee to confirm Jim's homosexual frolics. Jay Babcock, who started with PTL as a gardener and rose to be its $34,000-a-year show producer, told a grand jury about thirty sexual encounters with Jim over a three-year period. He later described them in detail in a *National Enquirer* article (October 24, 1991) titled "Jim Bakker Blackmailed Me to Be His Gay Lover."

According to Babcock, and to accounts in Shepard's book, the episodes involved each masturbating the other. Jim "never viewed his gay relationships as marital infidelities," Babcock said. "I think in Jim's mind, if he was having sex with men he wasn't being unfaithful to Tammy. . . . In his twisted mind Jim never considered himself gay. He viewed his dalliances with men as merely a way to relieve stress." Babcock, who insists he is not gay, said his antics with Jim "filled me with disgust and shame," but he knew his refusals would mean losing his job. He recalls asking Jim how he could afford so many expensive suits, "Because," Jim replied, "stupid people keep sending me money."

Bakker speaks often in his book about Jim Toms, his Hendersonville Christian friend and attorney. For a while, after his release from prison, Bakker worked for Toms, and even painted the facade of Toms's office building. Bakker fails to mention that in 1995 Toms confessed to having bilked more than $1 million from his clients. He has declared bankruptcy and been disbarred in North and South Carolina. When Bakker was interviewed by Hendersonville's *Times-News* in October 1996 he had no comment on his lawyer except to say that Toms and his wife were "two of the finest people I have ever met." On March 19, 1997, Jim Toms pleaded guilty in Hendersonville, North Carolina, to eighteen counts of embezzling $1.4 million from his former clients. "I have let down my clients, the bar, and the courts," he told the judge. "Words cannot fully express my remorse for the wrongs I have done." His sentence will be decided in May.

Bakker's most emotional chapters are about his discovery that Tammy was leaving him. He prints many of her letters telling how much she loved him, and his equally passionate replies. To this day he insists he is still in love with her. He writes that after long prayerful struggle he has finally managed to forgive both Tammy and Roe Messner, as well as his three most bitter enemies, Ankerberg, Swaggart, and Falwell. He knows God has forgiven him for his sin with Jessica and for his lavish lifestyle.

The book closes: "I had once thought that God had abandoned me. I thought that my days of ministering for the Lord were done. I thought that I would never preach again. I was wrong."

The preaching has already begun again. Among evangelicals, public repentance goes a long way to redeem the sinner, and a penitent preacher's following may even be augmented as he presents the spectacle of a fallen man overcoming his sins. In March 1996 Jim was the "host" at two services at a church in Lakeland, Florida, whose pastor was his close friend Karl Strader. Strader's son Daniel was sentenced two years earlier to forty-five years in prison for defrauding elderly residents of more than $3 million. Reverend Dortch has set up a legal defense fund.

Unlike her earlier autobiography, Tammy's new book, *Tammy* (Villard, 1996), mentions no coauthor, though her page of thanks opens with "I want to acknowledge above all the Lord Jesus Christ." The book's jacket, like the cover of *I Gotta Be Me*, is a photo of Tammy in her familiar heavy makeup. Comics have long made jokes about her false eyelashes, thick lipstick, and mascara. T-shirts smeared with paint were sold, with the legend "I ran into Tammy Faye in the mall."

A few years ago Tammy toned down her makeup, saying she looked too much like Bozo the Clown, but the Bozo makeup is now back. It has become so well known a showbiz trademark that she can't abandon it. She says she still wears it while she sleeps, and intends to be buried with it on.

Tammy's new book covers much the same ground as her previous autobiography. There is much talk of sobbing, though not as frequent as in *I Gotta Be Me*, where she weeps on almost every page. She tells of occasions when Jim would lie face down on the floor and weep uncontrollably. Both he and Tammy like to describe sobs in triplicate. "I wept and wept and wept," Jim writes, after learning that his wife of more than thirty years was leaving him. And on page 617 he writes that after hanging family pictures in his Hendersonville home, "I broke down and sobbed and sobbed and sobbed."

When Jim and Tammy were guests on *Nightline* in 1987, it got the highest rating ever for a TV talk show. Ted Koppel begged them not to wrap themselves in Biblical quotations and spiritual admonitions, but Tammy, holding her pet dog Snuggles, closed the show by saying to listeners, "God loves you." Koppel later told a reporter, "He's a con man and she's a con woman."

Tammy's book gains in interest when she reveals her many clashes with Jim throughout their stormy marriage. She had wanted a child. He did not. (They had a boy and a girl.) He became more obsessed with building Heritage USA than with her. For a while they separated and considered divorce. Amazingly, they joked about their quarrels on PTL shows, prompting a deluge of letters from viewers who were praying for their reconciliation.

Tammy only briefly mentions her addiction to over-the-counter and prescription drugs, including a strong tranquilizer. Jim recalls how she once awoke screaming, "There's an elephant in the room." On a plane trip to a Palm Springs medical center for treatment, Tammy hallucinated. She saw an orchestra playing and people dancing on the wings. She saw a cat on the wing. They had to restrain her from trying to open a door to get out. Tammy completed her detoxification at the Betty Ford Clinic.

Tammy has nothing good to say about Alan Dershowitz. She claims he attended Jim's trial for only one day. "All I remember about Alan is his frequent calls to me in Orlando, and the threatening letters he wrote telling me he had to have the thousands of dollars we owed him for that one day he was there." Tammy had no money. A friend finally gave her $25,000 to silence his demands. It was not as much as he wanted, but "enough to satisfy him for a while." Dershowitz had every right to collect his fee,

Tammy continues, "but it was one of the straws that helped break the camel's back for me."

Tammy is even harsher on Judge Potter, who sentenced fifty-year-old Jim to forty-five years. She describes him as openly scornful of Jim, once calling him in front of the jury a "little sawed-off runt." During the trial, she writes, he would close his eyes as if asleep, and stick fingers in his ears when he didn't want to hear something. "He would yawn and belch and wink at the jury."

Tammy has lost none of her dislike for Pat Robertson. When Jim went to prison, Pat said Jim should serve every day of his sentence. "I thought this was an extremely cruel and cold-blooded thing to say. It wasn't enough that Jim was buried; Robertson had to shovel more dirt on his grave." When Tammy tried to get back into television she found doors closed in her face. An appeal to Oral Roberts and his son was never answered. Other televangelists refused to speak to her.

Rupert Murdoch's Fox Television finally hired Tammy to do a slightly risque talk show with actor-comedian Jim J. Bullock. When he told her he was gay, Tammy says, she replied: "When I look at people, I do not see gay or straight. . . . I see a person that God loves, and that His son Jesus died on the cross for. So who am I to judge?"

The Jim J. and Tammy Faye Show ended, after what Tammy calls "sixty fun-filled shows," when she discovered she had colon cancer. Tammy had been bleeding for a year, but relied on prayer rather than see a doctor. "I prayed for a miracle," she told a reporter, "but God said no." The Lord had earlier said no, Tammy wrote in *I Gotta Be Me*, when she prayed for him to raise her pet dog Chi Chi from the dead.

The tumor has been successfully removed and the cancer is in remission. The Bakkers' daughter, Tammy Sue, now married and with two children, lives near Jim. She is singing and preaching around the country. Jamie, their other child, had a difficult time getting over a period of smoking, boozing, and drug experimenting, but is now recovered and on his own as a youth minister. Roe has been indicted by a Wichita, Kansas, court for concealing $400,000 when he declared bankruptcy. He had lost millions when PTL went broke. Roe is now free on bond and appealing a jail sentence of twenty-seven months. The back jacket of Tammy's book says that if you want to join the Tammy Faye Fan Club you can reach it through a post office box in Rancho Mirage, California, where she and Roe live.

Although both books are largely self-serving flimflam, there is not the slightest doubt that Jim and Tammy are sincere in their Pentecostal faith. They speak in tongues. God talks to them in their hearts and in their dreams. They believe in Satan, demons, eternal hell, and the impending return of Jesus. For them the Bible is God's holy word, not open to doubt or cultural analysis. Both Jim and Tammy have rated high on IQ tests. Their religious opinions are not so much stupid as they are childish. Most of what they imagine to be their sins are trivial compared to a sin they are incapable of recognizing—the sin of willful ignorance.

Jim's narrative bristles with references to books that gave him spiritual strength while he was in prison. Virtually all are by conservative Christians whose views are close to his own. I would be astonished to learn that Jim has ever read a book skeptical of fundamentalist Christianity. His and Tammy's minds are untouched by knowledge of science or modern Biblical criticism. It may never occur to Jim that God was telling him all along to stop struggling to save souls, to get an education, and to work at an honest job.

The unsinkable Tammy Faye is today as happy, perky, and funny as ever. Although her book blasts the weekly supermarket tabloids (she calls them "rags"), when she married Roe she allowed The *Globe*, one of the worst scandal sheets, to cover the wedding. On the recent talk shows I've seen she giggles more than she sobs. After thanking Jim for saying he is still in love with her, she adds, "I must have done something right."

ADDENDUM

Church builder Roe Messner, battling prostate cancer, was convicted in 1995 on five counts of bankruptcy fraud. In March of the following year a judge sentenced him to twenty-seven months in jail. In 1996 Tammy underwent surgery to remove a cancerous tumor from her colon.

Larry King interviewed Tammy in June 1996. She was in good spirits. Larry spotted what looked like a fabulous diamond ring on her hand. Tammy giggled and said it was a hundred dollar zircon. Her husband, then free on appeal, appeared at the end of King's television show to proclaim his innocence of all charges.

In December 1996 I wrote to Tammy, at her home in Rancho Mirage, California, to ask how I could join her fan club. I got back what seemed to be a letter handwritten in blue ink. For ten dollars, it said I could get a color photo of Tammy, a ten percent discount on all her merchandise, and a year's subscription to her newsletter. The printed letter was addressed to "Dear Martin," and signed "God bless you Martin." Accompanying the letter was a sheet advertising four audio tapes and two videos of Tammy singing a mixture of pop tunes and hymns. I could get all six for $60.

Shown here is the color photo I later received, signed to me and my wife. Tammy's first newsletter headed "Hi Martin and Charlotte," was punctuated with "ha! ha!s." She reports she is free of cancer, and now more careful of what she eats. "I have always preferred CHOCOLATE over food! Now we keep carrots and celery in the fridge for me to snack on instead of CANDY everywhere. NOT NEAR SO MUCH FUN! ha! ha!" The letter ends, "I love you and look forward to hearing from you again," signed "Love, Tammy Faye (or) Tam, that is what all my friends call me. ha!" A postscript lists two of her books that can be had, postage free and signed, for $12.95 each.

Jim Toms, Bakker's Hendersonville, N.C., attorney, was sentenced on May 12, 1997, to fifteen-and-a-half years in

Jim Bakker and Tammy Faye

prison for having embezzled more than a million dollars from his clients. He is currently serving time. As for Reverend Wesley Fletcher, I have been unable to find out if he is alive or dead.

Jim Bakker moved to Los Angeles to work in what is called The Dream Center, run by The Los Angeles Church, to minister to inner-city poor. There he "fell madly in love," as he put it, with Lori Beth Graham, of Phoenix. They were married in 1998. Lori is young, blonde, and beautiful. "Bells rang," Jim said when he first saw her, and at age fifty-eight he felt as if he were sixteen. Lori is a divorcee; a battered wife who was into booze and hard drugs, and had five abortions, before she was born again.

Jim returned to the Carolinas in 1999, he and his wife settling in a log cabin south of Charlotte. The cabin was provided free by a Pentecostal ministry called Morning Star Fellowship. Jim said he plans to preach and write books. He hasn't ruled out involvement with the PTL complex in Fort Mill, S.C., which has been closed since 1997. He said he is eager to get back into television preaching.

On December 22, 1998, Jim was Larry King's guest. He said he had written a book about the Second Coming. While in prison he studied Revelation "a word at a time," and concluded from its description of days before Armageddon that the earth will be hit by a giant comet or asteroid! The Antichrist, he told Larry, is now alive and on Earth. His book, *Prosperity and the Second Coming* written with Ken Abraham, was published by Thomas Nelson in 1998. The book warns America that the world as we know it is about to end with terrible destruction followed by the return of Jesus. Is Jim faking it to make money? The sad truth is that he believes everything he says.

MORTON COHEN'S
LIFE OF
LEWIS CARROLL

T he most surprising of many surprises in *Lewis Car-roll: A Biography* (Knopf/Macmillan, 1995), a splendid new life of Carroll by Morton Cohen, is the conjecture that the Reverend Charles Lutwidge Dodgson (Carroll's real name) was romantically in love with Alice Liddell. More-over, Cohen believes that Carroll actually hoped that someday he might marry her. Not that he ever formally pro-posed. More probably, Cohen contends, he approached Alice's parents with a suggestion that "perhaps in the future, if her [Alice's] affections for him did not diminish, he would be happy to propose."

We do know that Mrs. Liddell, wife of the Dean of Christ Church, Oxford, where Carroll taught mathematics, was suddenly impelled to burn all of Dodgson's letters to Alice and to forbid him for a time to see her three charming daughters. The pages in Dodgson's carefully kept diary

This review first appeared in *Nature*, January 11, 1996.

covering this painful period were destroyed by Dodgson's niece. And there were rumors in Oxford, as one letter writer put it, "that Dodgson has half gone out of his mind in consequence of having been refused by the real Alice."

Much has been speculated about Dodgson's passion for photographing naked little girls. Although he eventually tried to destroy all these photos—there had been unpleasant gossip about them—four pictures survived. Cohen reproduces all four. Dodgson insisted that such photographing was a pure, untarnished effort to capture the beauty and innocence of prepubescent female bodies. Is it possible he deceived himself?

Cohen thinks he did. In a remarkable chapter titled "The Fire Within," he argues that Dodgson had all the desires of a heterosexual man, but because of his Anglican faith, augmented by Victorian mores, he struggled all his life to conceal and repress such feelings. Cohen believes it was the frequency of erotic images in Dodgson's dreams and reveries that explain the many passages in his diary where he seeks God's help in overcoming what he considered deplorable sins.

No one knows more about Dodgson than Professor Cohen, now retired from teaching English literature at City College of New York.

He has edited two volumes of Dodgson's letters, and a book of essays about Dodgson. His skill and tact in covering all aspects of Dodgsons complex, enigmatic personality is impressive. The book is no hagiography. It covers all of Dodgson's foibles: his prudishness, his snobbishness, his fastidiousness, his amusing eccentricities, and his indifference to literature and events outside of England. All this is overshadowed, of course, by his immense creative powers.

Many aspects of Dodgson's writings and beliefs are of special interest to *Nature* readers. He was not a great mathematician, but his love for mathematics and logic was unbounded. One reason the Alice books are so relished by mathematicians is because they bristle with mathematical riddles, paradoxes, jokes, and witty word play. Two of his books are about original mathematical puzzles: *A Tangled Tale* and *Pillow Problems*. More serious mathematical works include *Euclid and His Modern Rivals* and *An Elementary Treatise on Determinants*. His interest in formal logic produced *Symbolic Logic* and *The Game of Logic*. In recent years the discovery of Dodgson's unpublished papers on logic show him well ahead of his time in recognizing aspects of logic far from trivial.

Dodgson's confusing paradox about three barbers, based on ambiguity over the relation of "if-then," aroused widespread controversy. "What Achilles Said to the Tortoise," a paper in *Mind*, raised a deep question about the ultimate validity of all logic and mathematics. Such reasoning rests on the premises of formal systems, but can we be certain the premises are valid? To justify them one must make more fundamental assumptions, and so on into the abyss of an infinite regress.

Dodgson was constantly printing leaflets and pamphlets about recreational mathematics and word games. Cohen surveys them all: original board games; games with cards, words, and numbers; cipher methods; a mnemonic system for memorizing dates and large numbers; divisibility rules, improved systems of voting; and rules for croquet, for tennis tournaments, for playing billiards on a circular table, and for rapidly determining a given date's day of the week.

Geologists and life scientists may be surprised to learn that Dodgson accepted an ancient earth and evolution as God's way of creating life. He inclined toward the "jump" theory proposed by British zoologist St. George Mivart, a Roman Catholic who was excommunicated for his heretical opinions. Cohen quotes from Dodgson's diary: "read the whole of Mivart's *Genesis of Species*, a most interesting and satisfactory book, showing as it does, the insufficiency of 'Natural Selection' *alone* to account for the universe, and its perfect compatibility with the creative and guiding power of God."

In many other ways Dodgson was theologically more liberal than most contemporary Anglicans. He totally rejected, for example, the doctrine of eternal punishment for the wicked. "If anyone urges 'then, to be consistent, you ought to grant the *possibility* that the Devil himself might repent and be forgiven,' " he wrote in a letter, "I reply 'and I *do* grant it!'"

Literature on Dodgson continues to proliferate. Cohen's book is only the latest and most definitive of many biographies. There is even a biography by Anne Clark of Alice Liddell. Carroll societies flourish in England, the United States, and Japan, each issuing a periodical. Unexpurgated editions of Dodgson's diary, in many volumes, are now being edited by Edward Wakeling.

The Romans used a white stone or piece of chalk to mark on a calendar a day that gave special pleasure. The Latin phrase "a white stone day" was equivalent to what now is called a red letter day. Whenever an event was specially memorable to Dodgson, as on April 25, 1856, when he first met Alice (she was then almost four) he would write in his diary, "I mark this day with a white stone." Throughout the world Carrollians will be marking with white stones the day they begin reading Morton Cohen's marvelous book.

LEWIS CARROLL, PHOTOGRAPHER

S ince Charles Lutwidge Dodgson, better known as Lewis Carroll, died exactly one hundred years ago, a vast number of books and articles have been written about his complex, elusive personality. No one living knows him better than Morton Cohen, English professor emeritus of the City University of New York. He earlier wrote a splendid biography of Carroll, and edited a two-volume collection of Carroll's letters. Now he has given us a strikingly beautiful volume, *Reflections in a Looking Glass* (Aperture, 1998), about Carroll as a skilled, pioneer photographer.

While teaching mathematics at Christ Church, Oxford, and writing his two immortal fantasies about Alice, Carroll became fascinated by the exciting, newly invented art of photography. In those days the art was in its infancy. Those who posed for pictures had to remain perfectly still

This review first appeared in the *Los Angeles Times Book Review*, December 6, 1998.

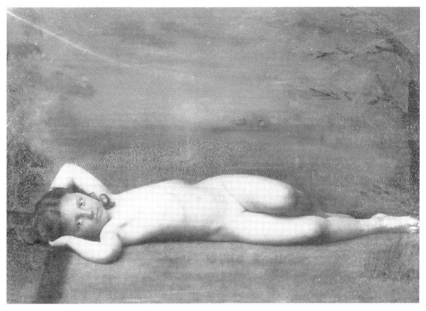

Carroll took some of his favorite photographs to be hand-colored by various female amateur artists. Four of Carroll's nude studies survive, all hand-colored. Above is Evelyn Hatch, 1879. From *Reflections in a Looking Glass*, Aperture, 1998.

for almost a minute, and the process of developing plates was tedious and difficult. Considering these primitive conditions, it is astonishing that Carroll's pictures are so wonderfully clear and detailed.

Professor Cohen introduces his selection of photos with a delightful summary of Carroll's life and achievements. He loved the theater, invented endless games and puzzles, enjoyed chess, liked to do magic tricks for children, and had a great fondness for art, music, and poetry. Although he took lessons in drawing, and even illustrated his first version of *Alice's Adventures in Wonderland*—it was a small book he hand-lettered as a gift to Alice Liddell —the art critic John Ruskin assured Carroll that his talent as an artist was mediocre. Carroll accepted this judgment. Nevertheless, he had remarkable feeling for composition that stood him well in his picture taking. Many of his landscapes and portraits are as artfully composed as a painting. He was unquestionably his century's finest photographer of children.

Although most of the photographs in this book are of Carroll's relatives, child friends, and children of friends, there are riveting portraits of such eminent Victorians as Alfred Lord Tennyson, Dante Gabriel Rosetti, novelist George MacDonald, painter Sir John Millais, and the actress Ellen Terry, one of Carroll's lifelong friends.

Only four of Carroll's many photographs of little girls in the nude have survived; hand-colored reproductions of all four are in Cohen's book. One of them, a frontal nude of Evelyn Hatch stretched out on a sofa, would probably be banned today as child pornography were it published as a *Playboy* centerfold.

What should one make of Carroll's fondness for photographing little girls *sans habillement*, as he once phrased it in a letter? A devout Anglican, Carroll was scrupulous about obtaining the consent of parents before he took nude photos. At the same time he was constantly distressed by gossip. Before he died, he destroyed almost all the negatives and prints he had retained of such pictures.

As for those who see impure motives in Carroll's nude photos, Professor Cohen has this to say:

> They see a dour, bleak, unhealthy world in these photographs, a repressed sexuality writ large. One wonders if they are not bringing their own neuroses to these works, twisting and despoiling their beauty. . . . Carroll was as successful in sublimating his emotional desires as he was in achieving his distinction as a photographic artist, and, certainly the two are related. But he was aware of himself and his unconventional desires, and he was honest and open with his child sitters and with their elders. . . . His photographs of children, even his nude studies, convey an innocence, an emotional purity characteristic of childhood.

Alice as a beggar girl, perhaps the most memorable photograph that Carroll ever took, ca. 1859. From *Reflections in a Looking Glass*, Aperture, 1998.

Lewis Carroll, Photographer

Carroll's best known photograph of Alice Liddell shows her garbed as a barefooted child beggar. Hauntingly lovely, and perhaps colored by Carroll himself, it graces the front cover of Cohen's book. Elsewhere Cohen has given his reasons for believing that Carroll was romantically in love with Alice, and at one time even hinted to her parents that when she became of age, would they consider a marriage proposal.

"He [Carroll] never married," Cohen writes, "but he probably wished to and would no doubt have made a loving husband and doting father had he done so. He claimed generally to be happy, but at least one observer guessed he was a 'lonely spirit and prone to sadness.'"

Professor Cohen ends his essay by quoting a typically Carrollian response that Carroll made to his artist friend Gertrude Thompson. She had suggested he sketch some nude children she had seen romping about on the seashore. It would be, Carroll told her in a letter, a

> . . . hopeless quest to try to make friends with any of the little nudities. . . . A *lady* might do it: but what would they think of a *gentleman* daring to address them! And then what an embarrassing thing it would be to *begin* an acquaintance with a naked little girl! What *could* one say to start the conversation? Perhaps a poetic quotation would be best. "And ye shall walk in silk attire." But would *that* do? . . . Or one *might* begin with Keats's charming lines, "Oh where are you going, with your love-locks flowing, And what have you got in your basket?" . . . Or a quotation from Cowper (slightly altered) might do. *His* lines are "The tear, that is wiped with a little address, May be followed perhaps by a smile." But *I* should have to quote it as "The tear, that is wiped with so little a dress"!

BOOKS AND
BOOKSHELVES

"As the reader's eye strays . . . from these pages," G. K. Chesterton wrote in the preface to his book *Tremendous Trifles*, "it probably alights on something, a bedpost or a lamppost, a window blind or a wall. It is a thousand to one that the reader is looking at something that he has never seen: that is, never realized." No writer ever tried harder than Chesterton to convey to his readers a sense of wonder and mystery about common, everyday objects.

Henry Petroski, who teaches history and civil engineering at Duke University, opens his latest book almost identically. "My reading chair faces my bookshelves, and I see them every time I look up from the page. When I say that I see them, I speak metaphorically . . . for how often do we really see what we look at?" Petroski shares

---

This review first appeared in *Civilization* (October/November 1999).

Chesterton's knack for looking at tremendous trifles with amazement and curiosity. That is how he saw bridges in his book *Engineers of Dreams*. That is how he saw the lowly pencil in *The Pencil*, and many other familiar objects in *The Evolution of Useful Things*.

The Book on the Bookshelf is a fascinating history of two related common objects, impeccably documented and beautifully illustrated. It is impossible to put this book down without seeing bookshelves as the culmination of a long, colorful, little-noticed history. Books had their beginnings in Greek and Roman scrolls; they were handwritten, mainly on papyrus, then rolled into cylinders, tied with string, and put in sleeves. Homer's *Iliad* would have required about a dozen scrolls.

Scrolls began to be replaced in about the second century by codices—folded sheets of papyrus sewn together and bound inside wood covers. In the Middle Ages, these were produced and preserved in monasteries—the only libraries of the time. Due to their great scarcity, books were generally kept locked in chests and, later, were chained to lecterns. Because chains were best attached to cover edges, chained books were shelved with their fore-edges outward. There was no need to see blank spines; volumes were often identified by patterns on their fore-edges.

After the great printing revolution, books were duplicated by the thousands, and chaining became burdensome. Printed books began to be shelved with spines outward, bearing the title, the author, and sometimes the date. Petroski entertainingly documents every aspect of shelving: the alleged Chinese invention of rotating bookcases; the need for bookends; how to prevent shelves from sagging; the rapid proliferation of bookshops; and a hundred other bits of bibliophilic minutiae.

Petroski discusses ways to preserve books from the baleful effects of sunlight and mildew, the damage done by dirty or saliva-wet fingertips, and the dreadful practice of dog-earing corners. An amusing appendix lists two dozen ways that shelved books can be classified—usually alphabetically by the last name of the author or title, or by subject, but also by such eccentric ordering principles as the order of publication or purchase, or by price or sentimental value. If one's purpose is to obfuscate, books might be classified by such opaque methods as alphabetically by the author's first name, or by opening or closing sentences.

Are books as we know them here to stay, or will the computer revolution radically alter them? Already entire "books" can be downloaded from the Internet. Future bookstores may carry books on microchips displayed on microshelves. Researchers at the Massachusetts Institute of Technology are currently working on a project they call "the last book." It looks like a traditional book, but contains electronic ink, or e-ink, which is applied to the page from within the book, instead of by a press. E-ink is composed of microscopic spheres, half black and half white, which flip over electronically to produce what looks like an ordinary page. Titles can be selected by pressing buttons on the spine. With such a "book," a reader may eventually be able to download any of the twenty million volumes now in the Library of Congress! Given the future outlook for the book itself, the bookshelf may well become a thing of the past.

There is not a dull page in Petroski's splendid treatise. It is impossible to put it down without seeing bookshelves in a fresh new light as the culmination of a long, colorful, astonishing, little known history.

TWO BOOKS ON CHRISTIAN SCIENCE AND MARY BAKER EDDY

A haunting photograph of Mary Baker Eddy, taken when she was a young widow, stares at you from the jacket and the frontispiece of Gillian Gill's splendid, detailed, impeccably researched biography, *Mary Baker Eddy* (Perseus, 1998). One stares back in fascination. Mrs. Eddy's gaunt face, especially her enormous eyes, seem tinged with suffering, perhaps also with a touch of madness?

What color were those eyes? Gill quotes from the third volume of Robert Peels's standard life of Mrs. Eddy: "Her eyes, which they (her students) described variously as blue, gray, violet, deep gentian, black, even brown. . . ." These clashing colors can be taken as symbols of how violently Mrs. Eddy's biographers differ in their accounts of her peculiar personality.

Eddy biographers are of two types. There are the

This is a much expanded and revised version of a review that first appeared in the *Los Angeles Times Book Review* (August 22, 1999).

hagiographers who see no defects in her character or in her divinely inspired revelations. On the other side are the hostile critics for whom Christian Science is neither Christian nor science. Their books portray Mrs. Eddy (I quote from Gill) as "shallow, egotistic, incapable of love; painted, bedizened, affected; hysteric, paranoiac, mad; ambitious, mercenary, tyrannical; man-eater, drug addict, mesmerist; illiterate, illogical, uncultured, plagiarist."

Gill's massive biography falls somewhere in between. She makes clear she is not a Christian Scientist, though we are not told what her religious opinions are. We learn from the book's jacket that she has a doctorate from Cambridge University, has taught at Wellesley, Yale, and Harvard, and has translated six French books about psychoanalysis and feminist philosophy. She is best known as the author of *Agatha Christie: The Woman and Her Mysteries* (1990).

Among the hostile biographers, Gill is especially harsh on Georgine Milmine, whose name appears on the first effort to cover Mrs. Eddy's life. It ran as fourteen installments in *McClure's Magazine*, followed by book publication in 1909 shortly before Mrs. Eddy died. As Gill makes clear, Milmine was merely a reporter who tracked down information. The articles actually were written by a young Willa Cather, then editor of *McClure's*. A devout Catholic, Cather had an understandably low opinion of Mrs. Eddy and her doctrines. When the University of Nebraska Press reprinted the book in 1993, Cather and Milmine share the title page.

Although Gill acknowledges a great debt to the Milmine/ Cather biography, she finds many of its assertions false or wildly exaggerated. Its lies, she insists, found their way into later hostile works, notably Edwin Dakin's *Mrs. Eddy* (1929), Ernest Baker and John Dittemore's *Mary Baker Eddy* (1932), Mark Twain's *Christian Science* (1907), and my own *The Healing Revelations of Mary Baker Eddy* (1993).

Gill refers to my book more than a dozen times, never favorably except for one footnote where she credits me with uncovering a bizarre event. After Mrs. Eddy died, a

Boston Christian Scientist acquired the property in Bow, New Hampshire, on which once stood the farm house where Mary grew up. He arranged for a huge granite replica of the Great Pyramid of Egypt to be erected on the field as a memorial to Mrs. Eddy who had once likened the permanence of her science to the permanence of Egypt's "Miracle in Stone." In 1962 the monument was mysteriously dynamited to bits. I later learned that the pyramid's destruction had been ordered, for reasons not yet known, by the church's Board of Directors.

Although Gill does not share Mrs. Eddy's reinterpretation of Christianity or her methods of healing, she writes that as she learned more about Mrs. Eddy "she came to command my respect, my admiration and even my affection. I wish, with surprising passion, that I could have met her. . . ." In writing about her, Gill goes on to say, she has tried to be fair to the prosecution. However, she has deliberately adopted the role of defense attorney by doing her best to offer counter arguments and let readers judge for themselves.

Throughout her long life Mrs. Eddy was constantly accused of having stolen her basic ideas from "Dr." Phineas Parkhurst Quimby, an amiable, almost illiterate faith healer in Portland, Maine. Quimby acquired a large following of patients convinced he had instantly healed them of every kind of illness known. He believed he had clairvoyant powers to diagnose ailments, and a paranormal ability to heal, even over long distances. All illness, he taught, was mental. For a while he practiced mesmerism, claiming that electrical energy flowed from his hands. He would dip his hands in water, press one hand on a patient's belly, then rub his or her head with the other hand. Later he discarded rubbing, believing that his cures resulted entirely from a patient's faith. He would simply talk to them, explaining that their pain was all in their head and it would vanish if they believed it would.

Mrs. Eddy suffered all her life from a variety of chronic ills, especially spinal discomfort and stomach disorders. In 1862, bed ridden with back pain, she traveled to Portland to see Quimby. In a long letter to a Portland paper she described her dramatic cure. (Gill quotes only a portion of this famous letter; you'll find the full text in the Milmine/Cather book and in my biography.) Here is how the letter ends:

> But now I can see dimly at first, and only as trees walking, the great principle which underlies Dr. Quimby's faith and works; and just in proportion to my right perception of truth is my recovery. This truth which he opposes to the error of giving intelligence to matter and placing pain where it never placed itself, if received understandingly, changes the currents of the system to their normal action; and the mechanism of the body goes on undisturbed. That this is a science capable of demonstration, becomes clear to the minds of those patients who reason upon the process of their cure. The truth which he establishes in the patient cures him (although he may be wholly unconscious thereof); and the body, which is full of light, is no longer in disease. At present I am too much in error to elucidate the truth, and can touch only the keynote for the master hand to wake the harmony. May it be in essays, instead of notes! say I. After all, this is a very spiritual doctrine; but the eternal years of God are with it, and it must stand firm as the rock of ages. And to many a poor sufferer may it be found, as by me, "the shadow of a great rock in a weary land."

Mrs. Eddy and Quimby became good friends. She visited him often to discuss his methods and his belief that pain and disease are not real. When he died four years after her miraculous cure, she published a poem with the wonderful title: "Lines on the Death of Dr. P. P. Quimby, Who Healed With the Truth that Christ Taught in Contradistinction to All Isms." The last of its four stanzas is as follows:

> Heaven but the happiness of that calm soul,
>   Growing in Stature to the throne of God;
> Rest should reward him who hath Made us whole,
>   Seeking, though tremblers, where his footsteps trod.

Mrs. Eddy fancied herself a poet and loved to see her doggerel published in newspapers and magazines. There are two book collections of her poems, several of which were set to music as Christian Science hymns. I was pleased to note that Gill momentarily abandoned her role as defense attorney to admit that all of Mrs. Eddy's poetry is "dreadful," and that one of her published short stories is "unredeemingly awful."

Gill agrees that Quimby had a profound influence on Mrs. Eddy, but then she shifts to her defensive mode to write many pages attacking the honesty of "Quimbyites" who stress Eddy's debt to Quimby. "As I shall show," she writes, "in the course of this book, the evidence that Mary Baker Eddy's healing theology was based to any large extent on the Quimby manuscripts is not only weak but largely rigged."

I do not think it was rigged. As I show in my book on Mrs. Eddy, evidence of the debt is monumental. Fleta Campbell Springer, in her excellent biography of Mrs. Eddy, *According to the Flesh* (1930), said it well: "Mrs. Eddy's denial of Quimby delivered a wound to her emotional body from which she did not recover. . . . The ghost of Phineas P. Quimby haunted her all her life."

Two weeks after Quimby died, Mrs. Eddy slipped on an icy sidewalk in Lynn, Massachusetts and, as she believed, severely crippled her spine. A homeopathic physician told her, she claimed, that her injury was incurable. (He later denied making such a remark.) Mrs. Eddy opened her Bible at random. Her eyes fell on Matthew 9:2, which tells how Jesus healed a man of palsy. As she said in a letter, "on the third day I rose from my bed" completely whole again.

Mrs. Eddy regarded this moment as the date of her discovery of Christian Science. She was then forty-five. In an autobiographical essay, she wrote that for twenty years she had been "trying to trace all physical effects to a mental cause," but not until this miraculous healing of her spine did she gain "scientific certainty that all causation was

Mind, and every effect a mental phenomena." By "every effect" she did not mean just effects on the body, but all that happens in the universe. Mind or God is everything. All else, like the Maya of Hinduism, is illusion. She likens this insight to Newton's discovery of gravitation when he saw an apple fall, forgetting that many years before she had applied the same metaphor to Quimby's teaching.

Another major charge that critics have hurled at Mrs. Eddy is that not only did she plagiarize ideas from Quimby, she had a habit of plagiarizing sentences from other writers. An anonymous article that first appeared in a magazine and later was reprinted twice without a byline in *The Christian Science Journal*, turned up in Mrs. Eddy's *Miscellaneous Writings* as her own. Gill considers this a "small sin," suggesting in a footnote that maybe Mrs. Eddy was the actual author.

Many uncontested instances of Mrs. Eddy's habit of copying without credit are not mentioned by Gill. One of Mrs. Eddy's treasured books was an anthology titled *Philosophic Nuggets: Bits from Rich Mines*. In my book I list a dozen nuggets that Mary purloined. Carlyle writes about a "time-world" that "only flutters as an unreal shadow." Mrs. Eddy writes in *Miscellany*: "This time world flutters in my thought as an unreal shadow." Ruskin: "A little group of wise hearts is better than a wilderness of fools." Mrs. Eddy, in *Miscellany*: "A small group of wise thinkers is better than a wilderness of drunkards."

Scores of other aphorisms by famous writers are repeated with only trivial changes by Mrs. Eddy, with no hint that the words are not her own. Some of such stolen passages remain in today's edition of *Science and Health*. In Mrs. Eddy's *Miscellaneous Writings*, a "Message to the First Members" contains paragraphs copied almost word for word from a sermon titled "The Man of Integrity," by a Scottish minister. It had earlier appeared in a British work. Gill does not mention this obvious plagiarism; you'll find the parallel passages in my book.

Mrs. Eddy kept a scrapbook in which she pasted newspaper columns with such headings as "Gems of Truth" and "Dewdrops of Wisdom." A raft of these dewdrops dropped into Mrs. Eddy's writings. Gill is also silent on this compulsive borrowing. Ironically, Mrs. Eddy was furious when she discovered that anyone had copied anything from *her*. Her third husband, Asa Eddy, in the third edition of *Science and Health* wrote (in Gill's words), "an indignant preface, fulminating against those who dared to plagiarize his wife's copyrighted work."

Although not exactly plagiarizing, the most glaring instance of Mrs. Eddy taking credit for words written by someone else brings us to the strange way in which *Science and Health* was heavily revised by Reverend James Henry Wiggin.

The first edition of *Science and Health*, the holy "Bible" of Christian Science, was privately financed by Mrs. Eddy in 1875. Never proofread, this tome of 456 pages had hundreds of typographical errors, not to mention Mrs. Eddy's mistakes in spelling, punctuation, and grammar. She had no sense of how to paragraph. The book is a chaotic patchwork of topics, constantly repeated, at times incoherent, that alter as abruptly as images in a dream. Only a few hundred of the book's thousand copies were sold.

This first edition was so badly written, swarming with so many outrageous assertions, that when its copyright expired in 1971 Christian Science officials actually managed, through the help of President Nixon's two Christian Science aides, H. R. Haldeman and John P. Erlichman, to persuade Congress to pass a bill extending the book's copyright another seventy-five years to keep any publisher from reprinting the edition. This scandalous effort at suppression is not mentioned by Gill.

After the courts ruled the bill unconstitutional, a facsimile was offset and distributed by the Rare Book Company of Freehold, New Jersey. On page 350 you can read Mrs. Eddy's claim of being able, like Quimby's so-called angel visits, to heal people at a distance. On page 400 you can find Mrs. Eddy's permission to let surgeons set broken bones, but the time will soon come, she adds, when this will not be necessary because "mind alone will adjust joints and broken bones." So confused is this first edition, Gill admits, that "no casual reader would have been likely to get past page 1." Then she shifts to her defensive mode and writes: "In my view, the 1875 edition failed because of the ignorance and stupidity of its public, not of its author." The book has many faults, she later says, but bears the imprint of a "brilliant mind."

Although Gill is reluctant to attack the many preposterous claims in the first edition of *Science and Health*, she is good in covering the book's endless changes. For the 1886 edition Mrs. Eddy hired Wiggin to undertake a monumental revision. Gill describes him as "a fat, jolly gentleman" who had resigned as a Unitarian minister to earn a "living as a writer and literary hack." He cleaned up Mrs. Eddy's lapses in spelling, punctuation, and grammar. Quotations from famous writers were added as epigraphs to the chapters. Scores of absurd passages were deleted. Bits of verse were added here and there. A twenty-page sermon titled "Wayside Hints" was entirely the work of Wiggin. Mrs. Eddy later removed it from her book.

For four years Wiggin struggled, as he put it, "to keep her [Mrs. Eddy] from making herself ridiculous." He also edited other books by Mrs. Eddy, and even tried to improve some of her poems. When their temperaments began to clash, Mrs. Eddy fired him.

In a long letter which Gill quotes only in part, Wiggin gave his private opinion of Christian Science and its founder. He calls it an "ignorant revival of one form of gnosticism." It teaches that "reality is a dream . . . that sin is nonexistent because God can *behold* no evil." Mary Baker Eddy is described as a "smart woman, acute, shrewd, but not well read, nor in any way learned." Many Christian Scientists, he writes, prefer early editions of *Sci-*

ence and Health because they sound more like Mrs. Eddy. "The truth is she does not care to have her paragraphs clear, and delights in so expressing herself that her words have various readings and meanings. Really, that is one of the tricks of the trade. . . . There is nothing really to understand in *Science and Health* except that God is all and yet there is no God in matter! . . . Matter and disease are like dreams, having no existence."

Gill grants that in this letter Wiggin scored some good points, but then she reverts to her defensive role to castigate him for accepting payments from Mrs. Eddy then later bad-mouthing her. His letter, she says, smacks of "sour grapes." In a footnote she agrees with a biographer that he was a "cad and hypocrite."

After Wiggin died, *Science and Health* was continually revised by Mrs. Eddy and others. Endless changes were made, large sections removed, new sections added, chapters shifted about, and hundreds of sentences altered. Not until 1910, a year after Mrs. Eddy's death, was the 226th edition frozen in its present form.

Gill is both entertaining and accurate in providing details about Mrs. Eddy's discombobulated life. She was born in 1821 on a farm in Bow, New Hampshire, to a poor family of devout Congregationalists. There were many quarrels with her father about his fundamentalist views on hell and predestination. Gill minimizes evidence that as a child Mary was subject to hysterical "fits" of temper. She attributes Mary's extreme thinness as a youth to crank diets and anorexia. As a child Mary believed she had psychic powers of clairvoyance. She claimed that once God called her name aloud three times and each time levitated her a foot above her bed.

Mary's first husband, whom she married at age twenty-two, was George Washington Glover, a Charleston, South Carolina, building contractor. He died of yellow fever six months after their marriage. Mary's only child, George Glover, was born a few months later.

Biographers differ over whether Mrs. Glover was gen-
uinely fond of her son, or only pretended to be. At age four
his care was taken over by a family nurse. Did Mary Baker
Glover send him away, or was he "taken away" as Mrs. Eddy
later said in her memoirs? "I never saw him again," she
writes, "until he reached the age of thirty-four and came to
visit me in Boston." In his mother's old age, George, eager to
get his hands on a fortune of several million, tried to have
Mrs. Eddy declared mentally incompetent. The court
rejected his claim after Mrs. Eddy showed up as sharp as a
tack. Gill defends the view, against all evidence, that Mrs.
Eddy "was a normal mother deeply attached to her son."

Mary's second marriage was to Daniel Patterson, a
good-looking dentist who practiced homeopathy on the
side. In her younger years Mary believed that homeopathic
drugs, diluted to a point where only a few or no molecules
remain, were superior to the harmful drugs of "allopathy,"
the homeopathic term for medical science. She often gave
homeopathic remedies to her patients as well as herself
until she became convinced it was solely faith in the drugs
(i.e., the placebo effect) that made them effective. After
twenty years of marriage, Dr. Patterson was said to have
left her for a wealthy married woman. At any rate, Mary Patterson divorced him in 1873 on grounds of desertion. He
died decades later in a poorhouse.

At age fifty-six Mary Baker married Asa Gilbert Eddy, a
mild little man ten years her junior. He had been a Spiritualist in his younger years, but after converting to Christian
Science became its first practitioner. When he died six
years later of heart failure, one of the craziest episodes in
Mrs. Eddy's life occurred. Mrs. Eddy refused to believe the
cause was a diseased heart. She ordered an autopsy, firmly
convinced that her husband had been killed by arsenic
poison transmitted through the air by the Malicious
Animal Magnetism (MAM) of an enemy.

All her adult life Mrs. Eddy believed in MAM. In her
mind this was an evil form of psychokinesis similar to the

practice of voodoo witch doctors who stick pins in dolls. This malicious energy could be sent over long distances to cause great harm to persons. To convince Mrs. Eddy that her husband had indeed died of heart failure, the "allopathic" surgeon brought her Asa Eddy's heart on a platter to show her the diseased valves. Mrs. Eddy never changed her opinion. She even wanted to add a new chapter to *Science and Health* about the true cause of her husband's death, but Wiggin talked her out of it on the grounds that the physicians involved would probably sue her for libel.

When Mrs. Eddy was nearing seventy she did another astonishing thing. She adopted as her son a forty-one-year-old former homeopathic physician, Ebenezer Foster who had become a Christian Scientist and who professed to adore her. One of his letters opens: "Dearest Sweetest Loveliest Mother of this World." Mrs. Eddy signed one of her letters to him: "Love love love, 0000000, Ever Thine Own." Soon Benny, as Mrs. Eddy called him, was caught juggling account books and enjoying an affair with his married private secretary. As usual, Mrs. Eddy blamed it all on MAM and booted him out of her life.

Mrs. Eddy was as unlucky in her friendships as in her husbands. Over and over again she would develop a warm, loving relationship with one of her devotees, only to have it turn into bitter enmity. When Mrs. Eddy was forty-eight she shared an office in Lynn with a twenty-one-year-old student, Richard Kennedy. "Was there ever a more glorious nature, a more noble soul, than Richard Kennedy possesses?" she wrote in a letter. When he moved away to practice healing on his own, Mrs. Eddy's fury was unbounded. In chapter 8 of the first edition of *Science and Health*, without mentioning his name, she brands Kennedy a "head-rubbing" quack, skilled in MAM. MAM, she adds, is "more subtle than all the other beasts of the field, it coils itself about the sleeper, fastens its fangs of innocence, and kills in the dark."

In the second edition of *Science and Health*, chapter 8 was replaced by a chapter titled "Demonology," in which

Mrs. Eddy continues to accuse Kennedy of trying to murder her with MAM. In the third edition her attacks on Kennedy's "gigantic evil of character" are even more inflammatory. Kennedy never married (Gill thinks he was gay), became an Episcopalian, and died in a mental hospital.

The third edition of *Science and Health*'s chapter on demonology swarms with vicious attacks on another of Mrs. Eddy's former associates, Daniel Spofford, who, like Kennedy, had dared to set up an independent practice. Spofford was the publisher of her book's second edition. Soon the former friends were quarreling over finances, and Mrs. Eddy became convinced as always that he was trying to injure her with MAM. She calls him a "criminal mental marauder that would blot out the sunshine of earth . . . and murder in secret the innocent. . . ."

In 1878 Spofford was excommunicated. Mrs. Eddy sued him twice, once for back tuition, once for using MAM to inflict great suffering on one of her female students. The second trial was known as the Ipswich Witchcraft trial because the injured student lived in Ipswich and the trial was held in Salem. Both suits were dismissed.

It was that same year that the most comic, most mystifying episode in Mrs. Eddy's life took place. Her third husband, Asa Eddy, and one of her students, were accused of plotting to assassinate Spofford! Spofford got wind of the plot and vanished. A rumor had it that his body had been found in a morgue. Mr. Eddy and his friend were arrested. Newspapers had a field day. The case against them was dismissed when Spofford turned up alive and well, and a witness confessed to giving false testimony.

The Christian Science cult, in spite of constant attacks, expanded with amazing rapidity. It became a prosperous, well-organized empire, with elegant churches throughout the land and abroad. It published impressive books and periodicals. Over this growing empire Mrs. Eddy ruled shrewdly and with an iron fist. There were endless lawsuits, many started by Mrs. Eddy, others by her enemies.

To her followers she became a saint. To mainline Christians and skeptics she became an object of ridicule. A town burned her in effigy. Gill spares few of the sordid details. Relationships of love never ceased to turn to hate. Mrs. Josephine Curtis Woodbury, a Boston society woman who became an ardent Christian Science leader, attracted a large following in the Boston area. Suddenly, to everyone's amazement, Mrs. Woodbury announced that she was giving birth to a child who had no father! She or Mrs. Eddy called it an "immaculate conception," misusing the Catholic term which stands, of course, for the Virgin Mary's freedom from original sin. Mrs. Woodbury named her son Prince of Peace, and said he was destined to redeem the world.

Most Christian Scientists are not aware that in her writings and in *Science and Health* Mrs. Eddy foresaw the time, in the far future, when marriages would cease and children could be born without sexual encounters. (For the relevant passages, see my biography of Mrs. Eddy.) The illegitimate child's father, as Gill reveals, was a young Christian Scientist in Montreal with whom Mrs. Woodbury had a brief affair.

When the Prince of Peace was baptized in a pool on Mrs. Woodbury's Maine estate, she reported he was under the water three times, smiling with open eyes. Gill does not mention that Mrs. Woodbury was alluding to one of the most notorious passages in *Science and Health*. In the book's final edition (pp. 556–57), it reads as follows:

> It is related that a father plunged his infant babe, only a few hours old, into the water for several minutes and repeated this operation daily, until the child could remain under water twenty minutes, moving and playing without harm, like a fish. Parents should remember this, and learn how to develop their children properly on dry land.

Mrs. Eddy never bought the immaculate conception story. Woodbury was excommunicated. Mrs. Eddy later allowed her back in the fold, but after she became involved in several unpleasant lawsuits, she was "forever excom-

municated." Undaunted, Mrs. Woodbury tried unsuccessfully to start a rival church in Boston. In a magazine article
she bashed Mrs. Eddy (if I may quote from my biography)
for "childish writing, her refusal to credit Quimby with the
origin of Christian Science, her unwillingness to admit she
had once been a Spiritualist medium, her 'demophobia,'
her love of money, her constant lying about her sensational cases of healing, forbidding her followers to read
anything except the Bible and *Science and Health*, her
autocratic control of the church, and in general for 'per-
verting and prostituting' the true healing!"

An angry Mrs. Eddy responded with a fiery speech in
which, without mentioning Woodbury's name, she likened her
to the Book of Revelation's whore of Babylon. When a Boston
newspaper printed the speech, Woodbury sued the paper for
libel. The case was dismissed after Mrs. Eddy swore she was
attacking only a type of woman, though everyone knew otherwise. In 1909 Woodbury published *Quimbyism: Or the
Paternity of Christian Science. Echoes*, an earlier book, was
a collection of poems almost as bad as Mrs. Eddy's. Wiggins,
of all people, supplied an effusive introduction that prompts
Gill to speak of "how low Wiggin had sunk."

The case of Mrs. Augusta Emma Stetson is almost as
funny as the Woodbury case. An attractive, buxom woman,
skilled in oratory, singing, and organ playing, Mrs. Stetson
began her career as an adoring disciple of Mrs. Eddy. She
showered Mother, as Mrs. Eddy was then called, with
costly gifts that included an ermine cape and a diamond
brooch. Her letters to Mrs. Eddy always opened with "My
Precious Leader," and ended with "Your loving child."

In one of her many thank-you notes Mrs. Eddy wrote:

> I am in great need of summer suits of clothing, will you
> send me samples of these? Oh how good you are to me:
> What can I do to pay you, tell me, dearest one? You are all
> the student that I can depend upon to clothe me, and
> inasmuch as you have done it unto me, ye have done it
> unto the Father.

Mrs. Eddy was, of course, quoting Jesus.

Hostilities quickly surfaced after Mrs. Stetson raised funds to construct New York City's First Church of Christ, Scientist, a huge, marble temple at Central Park West and 96th Street. Aware that Mrs. Stetson was becoming her greatest rival, Mrs. Eddy began to warn students that Stetson was beginning to use MAM against her. In 1909, shortly before Mrs. Eddy died, Stetson was excommunicated. She continued to write books about Christian Science (one ran to 1,277 pages!) without once being critical of her precious Mother. Stetson believed, or pretended to believe, that only church officials, not Mrs. Eddy, had banished her.

Mary Baker Eddy spent her final years as a lonely, frail, paranoid, pain-wracked old lady in her mansion on Beacon Street, in Chestnut Hill, Boston, surrounded by worshipping students and her ever faithful servant Calvin Frye. The house was pervaded by MAM. For her periodic suffering Mrs. Eddy frequently took morphine. She believed her ills were caused by MAM heaved against her by enemies. Student "watchers" were assigned regular shifts to combat this MAM by beaming out positive energies. Mrs. Eddy believed they even had the power to control the weather.

Mrs. Eddy's private secretary Adam Dickey, in his *Memoirs of Mary Baker Eddy* (1927), reveals that a year before she died Mrs. Eddy begged him to tell the world that she was "mentally murdered." This plea is confirmed by an entry which Gill quotes from Calvin Frye's diary:

> Some week or more before our beloved Leader passed from us, she called me to her and asked me to promise her that I would tell her students it was malicious animal magnetism that was overcoming her and not a natural result from the beliefs incident to old age and its claims of limitation. She had hoped to demonstrate the way over old age but the malice and hate which poured in upon her thought left through the next friends suit in the New Hampshire courts, I believe was largely burden which seemed to undermine her vitality.

The poor soul died in 1909, at age eighty-nine, convinced that her impending death would occur because she had not been strong enough to ward off the MAM of her enemies.

Mrs. Eddy lived to see her metaphysics and healing methods, and her weird reinterpretations of the Bible, become the foundation of one of the nations most successful, most prosperous religious sects. In recent years it has been declining, its officials constantly squabbling, its revenues drastically reduced. Visit one of its services today and you'll find the auditorium half empty, with elderly women outnumbering men by a large margin. It seems likely that Mrs. Eddy's most lasting legacies will be *The Christian Science Monitor* and her feminist emphasis on God as Mother as well as Father.

It is not hard to understand both the rise of Christian Science and its fall. It came into being at a time when medical science was in its crude infancy. Many of its drugs and practices, blood letting for instance, did far more harm than good. The power of the placebo was only dimly appreciated, not to mention that many illnesses go away by themselves. Stirring testimonies to Christian Science healing, such as those in the back of *Science and Health*, undoubtedly took place, though one must be sketical of such wild claims as the dissolving of cataracts, or claims by the church that on several occasions Mrs. Eddy raised persons from the dead.

Today, faith healing has been taken over by Pentecostal televangelists such as Oral Roberts and his son, Pat Robertson, and Benny Hinn. Newspapers are now less reluctant to print horrendous accounts of the deaths of children who would have lived if their Christian Science parents had sought medical help. Gill tells of several such tragedies, though I was disappointed by her lack of moral outrage. There must have been thousands of unreported deaths of Christian Scientists and their children. Although today many Christian Scientists will seek medical help in emergencies, Mrs. Eddy never encouraged it. "The author never knew a patient who did not recover when the belief of the disease had gone," she writes in *Science and Health* (p. 377). Suppose a man drinks poison without knowing it is poison, and dies. He had no false belief, so what killed him? Mrs. Eddy answers in *Science and Health* (pp. 177–78): He dies because of the widespread belief of others that poison kills!

Still another reason for declining interest in Christian Science is the spread of alternative medicines such as acupuncture, reflexology, homeopathy, herbal remedies, therapeutic touch, psychic healing, and so on. An ignorant public, knowing nothing about scientific evidence, finds such alternative healing more exciting than chatting with a Christian Science practitioner.

In her preface Gill asks:

> What other woman in modern times has had this kind of vision and this kind of power to implement it? What other woman had established a religious movement of international stature? In what other religion or religious denomination is a woman's message offered next to that of Jesus Christ? What other woman's writing is written and read side by side with the Bible? What other woman dared to subtitle her own work "Key to the Scriptures"?

There is one answer to all these questions: Ellen Gould White.

Mrs. White was the founder and inspired prophetess of Seventh-day Adventism, a Protestant fundamentalist cult with hundreds of churches around the world and with more believers now than Christian Science. Ellen White provided her own "key to the Scriptures" in numerous books that are today, by all Adventists, "read side by side with the Bible." A recent biography of Mrs. White consists of six fat volumes. Another scholarly biography was published in 1998.

As Christian Science fades, Seventh-day Adventism grows, fueled mainly by fundamentalism's current obsession with an approaching Armageddon and the return of Jesus as the year 2000 arrives. Unlike Christian Scientists, Adventists have no quarrel with medical science. Indeed, they run many of the nation's finest hospitals. A Christian Science hospital is, of course, unthinkable.

Ellen White was born six years later than Mrs. Eddy, and died only five years later. It's a great pity that these two most remarkable women ever to start and lead a new Christian faith never met. Had they done so, you can be sure that each would have thought the other a pious, ignorant, demented humbug.

"Mary Baker Eddy's child, her Church, is dying." So opens the last part of *God's Perfect Child: Living and Dying in the Christian Science Church* (Henry Holt, 1999), by Caroline Fraser. Her book is the most powerful, most persuasive attack on Christian Science to have been written in this century.

Why is Christian Science dying? One reason surely is the increasing willingness of the media to publicize preventable deaths of Scientist children. Another reason is the increasing ability of mainline medicine to heal the sick and prevent needless suffering. A third factor is the growing competition of other religious cults and the popularity of alternative medicine. Fraser likens Christian Science to a "half-forgotten old character at a crowded party elbowed aside by dozens of other attractive, if wacky, nondenominational alternative-health-care faddists, herbalists, power-of-prayer healers, and peddlers of natural nostrums."

> The healers and self-helpers have been enormously Successful, churning out best-seller after best-seller and movement after movement in each successive generation: Napoleon Hill and his *Think and Grow Rich*, which promises wealth to those who could tap into "Infinite Intelligence"; Norman Vincent Peale's *The Power of Positive Thinking*; Werner Erhard's est; Dr. Joyce Brothers and her *How to Get Whatever You Want Out of Life*; Deepak

Two Books on Christian Science and Mary Baker Eddy

Chopra's ayurveda and his dizzying proliferation of books on how to become "ageless" and "timeless"; Bernie Siegal and his *Love, Medicine, and Miracles*; Marianne Williamson and her popularization of *The Course in Miracles*, a textbook fundamentally inspired by Christian Science; Louise Hay (a former Scientist) and her *You Can Heal Your Life*; Andrew Weil with his wise family-physician face, his guru's white beard, and his *Spontaneous Healing*. They all came out from under Mary Baker Eddy's overcoat.

Caroline Fraser is a young writer and poet with a Harvard doctorate in literature. She was raised by a stern, dedicated Christian Science father and a skeptical mother who did her best to avoid arousing her husband's anger by expressing doubts. If Caroline had a headache, her mother would secretly slip her an aspirin. Mr. Fraser had no doubts. The television set was turned down whenever a commercial for a medicine came on. He refused to wear a seat belt because it implied that a collision could occur. Their sailboat had no radio because he knew they would never have an accident that would require one.

At age four Caroline would gaze at tables that *looked* real, but actually were not. "They weren't even there," she writes. "They were matter, and matter was Error, and Error did not exist." In Sunday School she learned that although error doesn't exist, somehow it comes from Mortal Mind. "Mortal Mind was thinking we were hurt when we fell down. Mortal Mind was forgetting to go to the bathroom . . . and wetting our pants. . . . Mortal Mind was having a tantrum. Mortal Mind was crying. The hardest thing to understand about Mortal Mind was the fact that it didn't exist."

Michael Schram was a Sunday School friend whose parents divorced because Mr. Schram couldn't tolerate his wife's Christian Science beliefs. When Michael began having stomach pains and repeatedly vomiting, his mother sought only the help of a Scientist practitioner. After several days of great pain, Michael died. Not until the third day did his mother call a funeral home. She and her practi-

197

tioner had been praying over the decomposing corpse hoping to raise Michael from the dead. An autopsy revealed that Michael died from a ruptured appendix.

The case created a media uproar. The Church's local PR man blamed the boy's death on the fact that the practitioner wasn't listed by the church. "He did not tell the papers," Fraser adds, "that listed practitioners undergo only two weeks of religious training and have no way of recognizing the symptoms of a ruptured appendix or any other illness even if they believed such illnesses existed, which they dont." No charges were filed because, as the prosecutor said, there was no proof that the mother or her practitioner expected Michael to die.

Michael's avoidable death ended Caroline's faith in Christian Science. "I did not write this book," she says, "so that . . . children like Michael Schram will not have died in vain. Michael Schram *did* die in vain. So have dozens, or hundreds, or even thousands of other Christian Science children; no one will ever know how many. They died for nothing. They died for an idea."

The first third of Fraser's book is a skillful account of Mary Baker Eddy's deluded, discombobulated life. The rest of the book is about the church she founded, its meteoric rise, its influence on American culture, and its recent rapid decline. No one has written more entertainingly and accurately than Fraser about the history of Christian Science after Mrs. Eddy died in 1910. No one has more colorfully covered the Church's endless bitter schisms and bad judgments that have dogged the Church and in recent years almost plunged it into bankruptcy.

In a section on Christian Science in Hollywood, Fraser reveals how fashionable Christian Science became among movie notables in the 1930s and 1940s. Believers included Cecil B. DeMille, Joan Crawford, Mary Pickford, Ginger Rogers, Doris Day, Robert Duvall, George Hamilton, Mickey Rooney, and a raft of others. Irving Schuman's biography of Jean Harlow charged that her death resulted from the fanatical beliefs of her Scientist mother. This rumor inspired a forgettable film in which Ginger Rogers took the evil mother's role. In her autobiography Ginger tells of many miracle healings, including her own recovery from pneumonia while she was appearing onstage in *Mame*. She never missed a performance. Hollywood had early plans to make a movie about Mary Baker Eddy, starring Mary Pickford, but the Church raised such a howl of protest that the film was never made.

In recent years, Fraser tells us, Carol Channing has been the most open in her praise of Christian Science. "I'm a Christian Scientist and we have no such thing as age . . . we don't believe in birthdays. Now, isn't that a nice religion to have?" Channing has never missed a performance even though she once finished a show after falling on stage and breaking an arm.

President Nixon's two famous assistants, H. R. Haldeman and John Erlichman, were, as I said earlier, devout Scientists. Fraser explains how they managed to persuade Congress to pass a law exempting Mrs. Eddy's *Science and Health* from becoming public domain when its copyright expired, a provision made for no other book. The Church did not want to lose control over the book's flawed, almost illiterate first edition. The law was soon tossed out by a court of appeals. Haldeman's son Peter, in a memoir published in the *New York Times Magazine*, tells how his father and mother refused all medical help that might have relieved his father's suffering and perhaps prevented his death from untreated cancer.

The closing chapters of Fraser's eye-opening book detail the hard times that have fallen on the Church in recent years. Consider, for example, the embarrassing case of Bliss Knapp. A Christian Science lecturer, Knapp became convinced that Mrs. Eddy was divine in the same sense that Jesus was divine. He defended this heresy in a work titled *The Destiny of Mother Church*. When he begged the Church to publish his book in 1948, the Board of Directors reacted in horror. They even wanted him to burn the plates he had used to make copies for the Library of Congress.

Then a funny thing happened. Knapp's widow and his sister-in-law, before they died, left a will in which they offered some $90 million from Knapp's estate to the Church provided it publish Knapp's book, stamp it "authorized," and place it in all reading rooms. Otherwise, the millions would go to Stanford University and the Los Angeles County Museum of Art. Again the Church said no.

"But in 1991," Fraser writes, "with the Church facing immanent bankruptcy, *Destiny* suddenly didn't seem so abhorrent." Arrangements were made with attorneys for Knapp's estate to revive the original offer. Over vehement protests from many Scientist bigwigs, the Church finally published *Destiny*. The directors denied this was done to get the much needed loot, but nobody believed them. The media picked up the scandal, comparing the Church to Judas. Many top officials resigned over the fracas.

Attorneys for Stanford and the art museum managed to block the bequest, then $97 million, on grounds that the Church had not met the terms of the will by calling the book "authorized literature." In 1993 six hundred reading rooms refused to carry the book. A settlement the following year gave the Church 53 percent of the cash, the rest to be split between Stanford and the museum. You can still buy what most Scientists consider a blaphemous book. "Authorized literature of the First Church of Christ, Scientist" is now printed on its copyright page.

The Church today remains in a state of great confusion and financial decay. Fraser ticks off the figures. *The Christian Science Monitor* has lost 100,000 subscribers since 1988, and only 25 percent of those who remain are Scientists. Branch churches are closing at a rate of 2 percent annually. The number of practitioners is declining at a rate of 5 to 6 percent a year. Church membership is estimated at less than 65,000.

The Church is now making a valiant effort to regain its

former glory. Virginia Harris, the Church's new energetic leader, is struggling to restore interest in *Science and Health* by promoting a new trade edition with ads that read "For People Who Aren't Afraid to Think." Fraser's profile of Harris is not flattering. "Unlike the Church's own lost children," she writes, "who have not been resurrected, Christian Science may well be." But she thinks and hopes it unlikely.

POSTSCRIPT

My review, even in its truncated form, naturally produced an angry response from a Christian Science spokesman. His long letter, and responses by Caroline Fraser and me, appeared in the *Los Angeles Times Book Review* (September 19, 1999), and are reprinted below:

> Martin Gardner's article "Mind Over Matter" purports to review three books: Mary Baker Eddy's *Science and Health With Key to the Scriptures*; Dr. Gillian Gill's 1998 biography, *Mary Baker Eddy*; and Caroline Fraser's new *God's Perfect Child: Living and Dying in the Christian Science Church*. Had these books received equal treatment along with their equal billing, the first two books would have considerably redressed the pointed imbalance of the latter.
>
> Clearly the reviewer favors Fraser's rancorous indictment of Eddy and her church over the scholarly study by Gill. As to the article's "review" of *Science and Health*, Los Angeles Times readers would have been hard-pressed to find one. The contents of the final edition of *Science and Health*—the culmination of Eddy's refinement of her ideas in the book over several decades—are barely mentioned. Instead Gardner focuses on the book's first edition. Though this edition is of genuine historical interest, there were six major revisions that followed.
>
> The story of *Science and Health*, selling more than 100,000 copies annually in sixteen languages, will continue to be told in lives healed and transformed. To the

many *Los Angeles Times* readers searching for deeper spirituality and prayer that brings solutions, a reading of *Science and Health* would be of great interest. Certainly, the 42 percent of Americans who use alternative treatments would be amazed at the review's reiteration of Fraser's dismissive attitude toward alternative and mind-body interventions, especially considering the nearly 300 studies documenting the benefits of prayer. Such a view lags sorrowfully behind the course curricula of 64 percent of American medical schools that now teach complementary and alternative medicine—the near majority of which feature courses in spirituality and healing.

Both Gardner and Fraser place undue reliance on an unreliable source, the 1909 biography of Eddy attributed to Georgine Milmine. Gardner justifiably praises Gill's *Mary Baker Eddy* as an "impeccably researched biography," yet his review inexplicably ignores the fact that in more than twenty references to the Milmine biography, Gill thoroughly discredits its sources and conclusions. (Gill's appendix documents her conclusive evidence against the propagandistic intent of the Milmine volume.)

The review's blanket statement, "To Christians and skeptics she was an object of ridicule," is easy to write but unsupported. Consider the words of Clara Barton, founder of the American Red Cross: "Love permeates all the teachings of this great woman [Mary Baker Eddy]. . . . Looking into her life history we see nothing but self-sacrifice and selflessness." And consider, too, that many evangelical Christians, including clergy, who have embraced this reinstatement of "primitive Christianity and its lost element of healing," also have recognized Mrs. Eddy as a Christian benefactor whom they deeply respected. Moving accounts of healings experienced through reading *Science and Health* fill that book's last chapter, including some by those who began as skeptics. Many Christians whose love of the Scriptures has been deepened by the book's spiritual insights into the Bible credit the book with restoring their faith, their moral dominion, and their health. Graciously, Gardner acknowledges these healings: "Stirring testimonies to Christian

Science healing, such as those in the back of *Science and Health*, undoubtedly took place. . . ."

These healings have occurred in great numbers and still do. Some 60,000 documented cures have been published by the church. Among children, Christian Science treatment has permanently healed such conditions as multiple head injuries, prenatal hydrocephalus, grand mal epilepsy, severely lacerated kidney, asthma, and mononucleosis, all medically diagnosed. To include this record would show the same sort of fairness routinely accorded medical advancements. So would an acknowledgment that Christian Scientists are always free to choose the healthcare system they deem best for themselves and their children.

Claims for spiritual cures need not escape scrutiny—but what kind of scrutiny? No one would think it fair to focus only on those cases medical doctors don't cure, or on medical mistakes alone, to assess what the practice of conventional medicine can achieve. Yet, in effect, this is what Gardner and Fraser do by zeroing in on the rare losses of children to Christian Science parents, losses as tragic as those experienced by parents whose children have received medical or other treatments. Those willing to consider the fuller record will see four and five generations of children whose parents have relied on Christian Science as their primary choice for health care, who have been well looked after, consistently healthy and regularly healed of illness and injuries through prayer to God alone.

Despite the article's repetition of notions such as the church's "fall" and "financial decay," its reserves are strong, and other than minor equipment leases, the church is free of debt. Further, despite being well into a major restoration of its international headquarters facilities, the church is devoting increased resources to meeting the growing public interest in the healing ideas of Christian Science. Demands on its international speakers bureau are increasing and recent years have seen wider distribution of *Science and Health* and other publications, including the *Christian Science Monitor*. Shortwave and domestic radio broadcasts continue to cover much of the world, and information on the church and its publications is available on the Internet.

Finally, what does it say about literary criticism in America today that any author—such as the one here, who never requested access to the church's historical archives—should be gratuitously praised for a personal attack on the religious faith and practice of fellow citizens, many of whom have served and continue to serve their country with distinction and make substantial contributions in their local communities?

It would be a shame if the biases in Gardner's review and in Fraser's book were to persuade anyone to ignore primary sources on Christian Science or to preclude them from deciding on its merits for themselves. Eddy invites the thinker into the pages of *Science and Health*. This book continues to speak for itself, leading spiritual seekers to satisfying answers about how to understand and draw closer to God—resulting in healing and individual well-being.

Gary A. Jones
Committees on Publication
The First Church of Christ, Scientist
Boston, Mass.

Martin Gardner replies:

I was not asked to review *Science and Health*, nor did I make any effort to do so. The title was added to the other two books by the editors (for readers interested in acquainting themselves with Mrs. Eddy's writings). If anyone is curious to know what I think of Mrs. Eddy's Bible, in all its endless revisions, they can buy my *Healing Revelations of Mary Baker Eddy*, soon to be reissued in paperback by Prometheus Books.

Gary Jones ticks off a list of horrible ailments that he claims have been completely cured by Christian Science. You don't have to believe him. Who diagnosed the ills and confirmed the cures? Christian Science practitioners? Incompetent doctors? One wonders how such awful ills could have gone away if, as Scientists believe, they never existed in the first place.

Yes, some of the cures described in the back of *Science and Health* may have been genuine. Many ailments go away without treatment; others respond to the power of faith in a treatment no matter how bizarre. Christian Science has a dismal record of not providing detailed confirmations of its more miraculous healings. The letters in *Science and Health*, describing such cures, are signed only with initials and without street addresses. This made it difficult to verify facts even at the time and certainly makes it impossible now.

There are claims, for example, that cataracts have been made to disappear. As any ophthalmologist will tell you, this is as impossible as unfrying an egg. There has never been a documented case of reversing the clouding of an eye lens. Patients with a strong belief that they have been cured of cataracts may fancy for a while they are seeing better, but that isn't confirmation of a cure.

Miraculous testimonials of healing are a dime a dozen in books by the most preposterous of cranks. A faith healer like Oral Roberts can claim he once brought back to life a dead child, but who believes him except other Pentecostals? I recently wrote in the *Skeptical Inquirer* two columns, one on a book about how to cure every ill known to humanity by rubbing the bottoms of feet, the other about a book making similar claims for drinking one's urine. Both books swarm with stirring testimonials of wonderful cures.

The most shocking statement in Jones's letter was his equating the deaths of Christian Science children, whose lives could have been saved by medical doctors, with children who die under orthodox medical treatment. That was a hit below the belt.

Christian Science is a non-Christian, nonscientific cult. Whenever such a cult is criticized, you can be certain it will arouse howls of angry protests from true believers. It's good that so few Christian Scientists have read my biography of Mrs. Eddy. They might have died of unreal apoplexy.

Caroline Fraser replies:

The letter by Gary Jones, spokesman for the Christian Science Church, misrepresents my scholarship in *God's Perfect Child: Living and Dying in the Christian Science Church*, claiming that I rely on an "unreliable source," the 1909 biography of Mary Baker Eddy written by Willa Cather and Georgine Milmine. More than 100 end notes in my book refer the reader not to the Cather-Milmine biography but to the church's own published source on Eddy's life, the three-volume biography by Robert Peel. Many other notes in my book refer to primary sources on Eddy and the history of Christian Science, not all of which are controlled by the church. As I recount in the book, it was the church's history of manipulating scholarship on Eddy that led me not to request access to their archives, a request that, in any case, would almost certainly have been denied.

Moreover, I took pains in my book to point out that Christian Scientists are productive members of society, describing the successful careers of Scientists from Ginger Rogers to Joseph Cornell, so Jones's wounded lament that I have maligned these good citizens is puzzling. I do criticize the overzealous religious practice of Christian Science parents whose medical neglect of conditions ranging from diabetes to meningitis has caused the suffering and deaths of their children, deaths which were not only "tragic," as Jones would have it, but, more to the point, unnecessary. Such conduct calls into question the moral authority of a church that has, throughout its history, rigidly pursued and enforced an extreme position against medical care that has indeed "transformed" lives but, in many cases, not for the better. Jones would like to paint my analysis as bigotry, but his defensive and inaccurate response is yet another example of the church's intolerance for constructive criticism.

THE MEME MEME

Paradigm Shift or Frivolous Fad?

T he word *meme* was introduced in an offhand way by Oxford University biologist Richard Dawkins at the end of his book *The Selfish Gene* (1976, revised 1989). Reviewers liked to quote Samuel Butler's remark that a chicken is an egg's way of making another egg. Dawkins's theme is similar. An organism is a gene's way of making other genes. The genes are "selfish" in the sense that they care not a fig about the welfare of the organisms that preserve and keep them going.

A meme (it rhymes with dream) is short for memetics, a word denoting mimicry. Dawkins wanted to identify a self-replicating unit of culture that would play a role in cultural evolution similar in some ways to the role played by self-replicating genes in the evolution of bodies. He chose the word meme because it had one syllable, like gene, and because it sounded like gene.

This review first appeared in the *Los Angeles Times*, March 5, 2000.

A meme is anything humans do or say that is not genetically determined, and is passed from person to person by imitation or copying. The term has already entered the *Oxford English Dictionary* where it is defined as "an element of a culture that may be considered to be passed on by nongenetic means, esp. imitation."

Memes are invisible self-replicators that live in human brains the way genes live in cells. They become tangible when they jump from one brain to another. Relativity theory, for example, slumbered in Einstein's brain as a meme before it went public in a technical paper.

Like genes, memes are selfish in being totally indifferent about their effects, for good or ill, on a culture. Like genes, useful or good memes tend to survive. Harmful memes usually die out. As Dawkins puts it, "A meme that made bodies run over cliffs would have a fate like that of a gene for making bodies run over cliffs. It would tend to be eliminated from the meme-pool." Although genes replicate much more accurately than memes, they do make copying errors, mutations that are usually harmful, but may aid the survival of a species or have no effect one way or the other. Memes likewise suffer from frequent copying errors that have similar effects on society.

Every idea or form of behavior that you learned from someone else, not on your own, is a meme. The list is endless: gestures, tunes, catch phrases like "Kilroy was here," fashions in clothes such as the current droopy trousers worn by young boys, ways to make anything (pots, chairs, cars, planes, skyscrapers), marriage customs, diet fads, art, novels, poems, plays, operas, tools, games, inventions, ideas in science, philosophy, and religion—all are memes.

What sociologists call mores and folkways are memes. From a "memes-eye view"—a favorite phrase of memeticists—all your beliefs about *anything* are clusters of memes. If you are a skeptic, your skepticism is made of memes. Was Jesus the Son of God? If you think yes, that's a meme. If you think no, that's also a meme. Belief or nonbelief in the importance of memes are memes.

There are three ways that memes leap from person to person. They jump vertically from parents to offsprings. They jump horizontally from person to person in ways resembling the spread of a virus. Thirdly, memes can move obliquely from someone to a near relative or friend. All words are memes. Spoken and written languages are clusters of memes.

After memes were tossed into the English language by Dawkins, memetics quickly caught fire, especially on the internet where it now has a cultlike following. Daniel Dennett, Tuft University's energetic and wide ranging philosopher, has become the top fugelman of memetics. He vigorously defends the meme concept at length in *Consciousness Explained* (1991), and at even greater length in *Darwin's Dangerous Idea* (1995). Two 1996 books promote the craze: Richard Brodie's *Viruses of the Mind: The New Science of the Meme*, and Aaron Lynch's *Thought Contagion: How Belief Spreads Through Society*.

This year a third spirited defense of memetics, Susan Jane Blackmore's *The Meme Machine*, with a high-praise foreword by Dawkins, was published by Oxford University Press.

Blackmore is a psychologist at the University of the West of England, in Bristol. Born in London in 1951, she is best known today among parapsychologists and skeptics for her disenchantment with psi research. It's a rare event when a believer in psi phenomena undergoes such a radical reversal! Blackmore's *Adventures of a Parapsychologist* (Prometheus Books 1986, revised 1996), is a charming account of her transition from psi believer to skeptic. She is also the author of two classic works: *Beyond the Body* (1982), a study of OBEs (out-of-body experiences), and *Dying to Live* (1993), a study of NDEs (near-death experiences).

Blackmore goes far beyond predecessors in her enthusiasm for memetics. She sees the meme concept as "a powerful idea, capable of transforming our understanding of the human mind." The ability to imitate one another is for Blackmore the main ability that distinguishes us from beasts. She grants that some garden birds have learned how to pry caps from milk bottles, monkeys have learned to wash sweet potatoes, and chimpanzees have discovered how to fish for termites with sticks. She denies, however, that such activities are true memes because they are individually learned, not passed around by imitation. Here she runs counter to the widespread opinion that chimps have a genuine low-grade culture, and that termite fishing is indeed passed along by copying.

In his forward to *The Meme Machine* Dawkins recognizes that Blackmore's vision of the future of memetics far exceeds his own. She has "greater courage and intellectual *chutzpah*," he writes, "than I have ever aspired to. . . ." Indeed, Blackmore speaks of her book as actually laying the foundation for a new science. Here are some of the big questions she believes can be answered fully only by invoking memes:

Why do we have such large brains? The conventional answer is that complex thinking has enormous survival value in the ability of a species to control its environment and keep itself from extinction. Blackmore puts it differently. Our brains are meme machines that have evolved to store and transmit memes. We are back to a version of Butler's topsy-turvy aphorism. Instead of brains creating memes, such as *Huckleberry Finn* and quantum mechanics, it is the other way around. "The enormous human brain," she writes, "has been created by the memes."

Many pages of *The Meme Machine* concern ways in which memes influence sexual behavior. All cultures have memes about what makes a person sexually attractive. Two popular such memes are that women should marry men taller and older than themselves. Such memes obviously influence mating choices, but Blackmore argues that we should override such trivial rules and mate with those who are the most skillful in "copying, using, and spreading memes."

Two chapters are devoted to a memetic theory of altruism. Animal altruism is now a hot topic among sociobiologists who are writing entire books to explain how unselfish behavior contributes to a species' survival. After an excellent survey of this literature, Blackmore tackles the topic in the language of memes. Instead of saying that altruism causes humans to behave unselfishly, she prefers the chicken and egg inversion. "Being kind, generous, and friendly plays an important role in spreading memes. . . ."

In her final chapter Blackmore follows Dennett in seeing consciousness and free will (two names for essentially the same thing) as illusions. They are "explained" by simply denying that they are real. For Blackmore and Dennett the notion of a "self" living inside our brain—an entity that makes decisions—is what Blackmore calls an "insidious and pervasive" notion created by the millions of memes that shuffle about inside our skull. There is, to put it bluntly, no such thing as a self:

> If I genuinely believe that there is no "I" inside, with free will and conscious deliberate choice, then how do I decide what to do? The answer is to have faith in the memetic view; to accept that the selection of genes and memes will determine the action and there is no need for an extra "me" to get involved. To live honestly, I must just get out of the way and allow decisions to make themselves.

Again:

> On this view, all human actions, whether conscious or not, come from complex interactions between memes, genes and all their products, in complicated environments. The self is not the initiator of actions, it does not "have" consciousness, and it does not "do" the deliberating. There is no truth in the idea of an inner self inside my body that controls the body and is conscious. Since this is false, so is the idea of my conscious self having free will.

Here is *The Meme Machine*'s final paragraph:

Memetics thus brings us to a new vision of how we might live our lives. We can carry on our lives as most people do, under the illusion that there is a persistent conscious self inside who is in charge, who is responsible for my action and who makes me me. Or we can live as human beings, body, brain, and memes, living out our lives as a complex interplay of replicators and environment, in the knowledge that that is all there is. Then we are no longer victims of the selfish selfplex. In this sense we can be truly free—not because we can rebel against the tyranny of the selfish replicators but because we know that there is no one to rebel.

Observe how this denial of a self meshes with the statement Butler made in jest. Instead of humans thinking about the world and their lives, exchanging information, inventing things, interacting with one another, experiencing pleasures and pains, the memetic language reverses everything. It's like bending over and looking at the world between your legs. You see the same things as before but from a different perspective. What is really going on is that billions of selfish memes are manipulating *us*. They have taken over our brains. They shape all our thoughts and actions. The memes are not our creations. We are theirs. We are just a meme's way of making other memes.

Because Blackmore shares Dennett's belief that all our decisions are determined by genes and memes—what we are wired to do by heredity and experience—she joins the ranks of thinkers known as determinists. One wonders what she would make of such famous arguments for free will as those found in William James's essay "The Dilemma of Determinism," or the attacks on determinism by later philosophers. It is easy to understand why Blackmore favors Buddhism over other faiths. For Dennett, she says, the illusion of self is benign; "for the Buddhist it is the root of human suffering."

Note the word *selfplex* in the last paragraph of Blackmore's book. It is her term for the cluster of genes and memes that give the illusion of a self. This brings us to the most serious objection that critics have hurled at memeticists. The notion of a meme is so broad, so fuzzy, so ill-defined, as to create more confusion than light.

A meme is supposed to be an element of imitation that can serve as a significant cultural unit. There have been a few earlier efforts to define such units. Blackmore cites two: "corpuscles of culture," proposed by anthropologist F. T. Cloak in 1975, and "culturgens," an equally ugly term suggested by physicist Charles Lumsden and biologist Edward Wilson in 1981. Neither has caught on. The question arises: Given that there are cultural elements called memes, how do we distinguish a single meme, such as the V for victory gesture, from a vast bundle of memes such as those that constitute a religion?

To answer this question memeticists have invented the word *memeplex* to denote a cluster of memes. Hilarious debates have raged over how to draw lines separating memes from memeplexes. For example, the first four notes of Beethoven's *Fifth Symphony*, da da da dum, as Blackmore records them, clearly are a meme because millions of people can hum those four notes without being able to hum the entire symphony. The symphony is, of course, also a meme, but best described as a memeplex because it consists of smaller memes.

The question of where to mark the boundaries along meme spectrums is not easy. "Laugh and the world laughs with you; weep and you weep alone." This clearly is a meme because so many people can repeat it without knowing they are the opening lines of "Solitude," a poem by Ella Wheeler Wilcox. All poems and novels are memeplexes. They often contain memorable lines that jump from person to person because they are so easily remembered. "Call me Ishmael," the first line of *Moby Dick*, is a meme. The first chapter of the novel is a memeplex, and the entire novel is a larger memeplex.

All of science is a memeplex, but it is hopeless to decide when a scientific assertion becomes small enough to be called an individual meme. Is the fact of evolution a meme or a memeplex? Roman Catholicism is a monstrous memeplex. What aspects of it deserve to be called memes? The doctrine of the Immaculate Conception? Or is this a memeplex made up of such memes as original sin and the Virgin Birth? Should an entire mass be called a meme or a memeplex? The gesture of crossing the heart is surely a meme, but we encounter great difficulty sorting out the memes that make up a doctrine as complicated as, say, the Atonement.

The point is that the notion of a meme is much too broad and ill defined to be useful in explaining human thinking and behavior. The memes-eye view is little more than a peculiar terminology for saying the obvious. Who can deny that cultures change in ways independent of genetics; ways involving information that is spread throughout society mainly by spoken and written words?

As Blackmore makes clear, memes have a physical basis of some sort inside brains where they are stored in one's memory in ways nobody understands. This is an important respect in which memes and genes differ. Genes have become visible. They are spots along the DNA helix that have been isolated and observed. They are as real as atoms. How memes live in brains is a mystery.

When memes jump from brain to brain they often are transported by what memeticists call meme vehicles. Obvious examples of such carriers are newspapers, periodicals, books, and recordings. Libraries, museums, and art galleries are huge vehicles for storing and passing along memes. "A scholar," writes Dennett, paraphrasing Butler, "is just a library's way of making another library."

All inventions are memes and memeplexes. Many also serve as vehicles. When you see a car, for example, you are seeing a vehicle carrying the meme of the wheel. Telephones, radios, television sets, movies, and computers linked by the Internet are twentieth-century vehicles for transmitting memes and memeplexes with speeds that would have been incomprehensible to earlier cultures.

For Dennett and Blackmore, memes offer profound new insights into human nature, and even lead to theories which may soon be testable. They see memetics as a science now in its infancy. To critics, who at the moment far outnumber true believers, memetics is no more than a cumbersome terminology for saying what everybody knows, and which can be more usefully said in the dull terminology of information transfer.

Stephen Jay Gould, in a 1996 debate with Blackmore, called memes "meaningless metaphors." Blackmore cites a letter in the *New Scientist* (February 12, 1994) in which British philosopher Mary Midgley calls memes "mythical entities" that are a "useless and essentially superstitious notion." H. Allen Ore, a University of Rochester geneticist, was quoted in *Time* (April 18, 1999) as dismissing memetics as "an utterly silly idea. It's just a cocktail party science."

Let's try a linguistic thought experiment. In all human cultures, even in chimp society, objects not connected to the body are shifted from place to place. Call every such move a *tran*, short for translocate or transfer. Moving our shoes when we walk, run, or dance are not trans because the objects are attached to our body. Nor are the movements of things in cars, trains, ships, planes, and elevators examples of trans because the propelling forces are independent of us even though we may direct such movements.

Examples of genuine trans abound. Pitched and batted baseballs are obvious trans, as well as the movement of objects in dozens of other sports: football, basketball, bowling, tennis, golf, hockey, pool, and so on. When a chess player pushes a pawn it's a tran. Dealing playing cards are trans. Raking leaves and moving vacuum cleaners and dust busters are trans. Serving food and washing dishes are trans. Hammering a nail and sawing wood are trans. Eating is a tran because food is moved from plate to mouth, though swallowing it is not because the food becomes joined to the body. Punching typewriter and computer keys are trans. Moving piano keys, trumpet valves, and drumsticks are trans. Digging ditches and cutting down trees are trans. Serving beer is a tran. There are tens of thousands of other examples.

A vexing question arises: How should we distinguish trans from transplexes? A pitched baseball is a tran, but if the ball is hit, caught, and tossed to first base, is that familiar sequence a tran or a transplex? Shall we call an entire inning a tran or a transplex? Should transplex be reserved for a complete game with its hundreds of trans?

What is gained by introducing the concept of a tran? Nothing. Trans are no more than a bizarre terminology for saying what is better said in ordinary language. We don't need a new science of tranetics to tell us that in every culture persons move things.

Are memes here to stay or will they prove to be as irrelevant as trans? Will memetics turn out to be a new science, or a harmless humbug destined to evaporate like Kurt Lewin's topological psychology that befuddled Gestalt psychologists in the 1930s, or catastrophe theory which two decades ago agitated a small group of overzealous mathematicians? Is memetics a misguided attempt on the part of behavioral scientists to imitate genetics with its gene units, and physics with its elementary particles? In a few years we may know.

WHAT IS MATHEMATICS, REALLY?

A physicist at M.I.T
Constructed a new T.O.E.*
  He was fit to be tied
  When he found it implied
That 7 + 4 = 3.

—Armand T. Ringer

W riting this review of Reuben Hersh's *What is Mathematics, Really?* (Oxford, 1997) has been an agonizing task; agonizing because I have such high respect for him as a mathematician and such low respect for his philosophy of mathematics.

Now retired, Hersh belongs to a very small group of modern mathematicians who strongly deny that mathe-

---

An edited version of this review appeared in the *Los Angeles Times Book Review*, October 12, 1997.
  *Theory of Everything

matical objects and theorems have any reality apart from human minds. In his words: mathematics is a "human activity, a social phenomenon, part of human culture, historically evolved, and intelligible only in a social context. I call this viewpoint 'humanist.'"

Later he writes: "mathematics is like money, war, or religion—not physical, not mental, but social." Again: "Social historic is all it [mathematics] needs to be. Forget foundations, forget immaterial, inhuman 'reality.'"

No one denies that mathematics is part of human culture. Everything people do is what people do. The statement would be utterly vacuous except that Hersh means much more than that. He denies that mathematics has *any* kind of reality independent of human minds. Astronomy is part of human culture, but stars are not. The deeper question is whether there is a sense in which mathematical objects can be said, like stars, to be independent of human minds.

Hersh grants that there may be aliens on other planets who do mathematics, but their math could be entirely different from ours. The "universality" of mathematics is a "myth." "If little green critters from Quasar X9 showed us their textbooks," Hersh thinks it doubtful that those books would contain the theorem that a circle's area is pi times the square of its radius. Mathematicians from Sirius might have no concept of infinity because this concept is entirely inside our skulls. It is as absurd, Hersh writes, to talk of extraterrestrial mathematics as to talk about extraterrestrial art or literature.

With few exceptions, mathematicians find these remarks incredible. If there are sentient beings in Andromeda who have eyes, how can they look up at the stars without thinking of infinity? How could they count stars, or pebbles, or themselves without realizing that $2 + 2 = 4$? How could they study a circle without discovering, if they had brains for it, that its area is πr^2?

Why does mathematics, obviously the work of human minds, have such astonishing applications to the physical world, even in theories as remote from human experience as relativity and quantum mechanics? The simplest answer is that the world "out there," not made by us, is not an undifferentiated fog. It contains supremely intricate and beautiful mathematical patterns from the structure of fields and their particles to the spiral shapes of galaxies. It takes enormous hubris to insist that these patterns have no mathematical properties until humans invent mathematics and apply it to the outside world.

Consider $2^{1398269} - 1$. Not until 1996 was this giant integer of 420,921 digits proved to be prime (an integer with no factors other than itself and 1). A realist does not hesitate to say that this number was prime before humans were around to call it prime, and it will continue to be prime if human culture vanishes. It would be found prime by any extraterrestrial culture with sufficiently powerful computers.

Social constructivists prefer a different language. Primality has no meaning apart from minds. Not until humans invented counting numbers, based on how units in the external world behave, was it possible for them them to assert that all integers are either prime or composite (not prime). In a sense, therefore, a computer did "discover" that $2^{1398269} - 1$ is prime, even though it is a number that wasn't "real" until it was socially constructed. All this is true, of course, but how much simpler to say it in the language of realism!

No realist thinks that abstract mathematical objects and theorems are floating around somewhere in space. Theists such as physicist Paul Dirac and astronomer James Jeans liked to anchor mathematics in the mind of a transcendent Great Mathematician, but one doesn't have to believe in God to assume, as almost all mathematicians do, that perfect circles and cubes have a strange kind of objective reality. They are more than just what Hersh calls part of the "shared consensus" of mathematicians.

To his credit, Hersh admits he is a "maverick" engaged in a "subversive attack" on mainstream math. He even provides an abundance of quotations from famous mathematicians—G. H. Hardy, Kurt Gödel, René Thom, Roger Penrose, and others—on how mathematical truths are discovered in much the same way that explorers discover rivers and mountains. He even quotes from my review, many years ago, of *The Mathematical Experience*, of which Hersh was a coauthor. I insisted then that two dinosaurs meeting two other dinosaurs made four of the beasts even though they didn't know it and no person was around to observe it.

A little girl makes a paper Moebius strip and tries to cut it in half. To her amazement, the result is one large band. What a bizarre use of language to say that she experimented on a structure existing only in the brains and writings of topologists! The paper model is clearly outside the girl's mind, as Hersh would of course agree. Why insist that its topological properties cannot also be "out there," inherent in what Aristotle would have called the "form" of the paper model? If a Hottentot made and cut a Moebius band he would find the same timeless property. And so would an alien in a distant galaxy.

The fact that the cosmos is so exquisitely structured mathematically is strong evidence for a sense in which mathematical properties predate humanity. Our minds create mathematical objects and theorems because we evolved in such a world and the ability to create and do mathematics had obvious survival value.

If mathematics is entirely a social construct, like traffic regulations and music, then it is folly to speak of theorems as true in any timeless sense. For this reason social-constructivists place great stress on the uncertainty of mathematics. Hersh doesn't mean something so trivial as that mathematicians often make mistakes. The fact you can blunder when you balance a checkbook doesn't falsify the laws of arithmetic. He means that no proof in mathematics can be absolutely certain. That $2 + 2 = 4$, he writes, is "doubtable" because "its negation is conceivable." No proof, no matter how rigorous, or how true the premises of the system in which it is proved, "yields absolutely certain conclusions." Such proofs, he adds, are "no more objective than esthetic judgments in art and music."

I find it astonishing that a good mathematician would so misunderstand the nature of proof. Benjamin Peirce, the father of Charles, defined mathematics as "the science which draws necessary conclusions," a statement his son was fond of quoting. Only in mathematics (and formal logic) are proofs absolutely certain. To say that 2 + 2 = 4 is like saying there are twelve eggs in a dozen. Changing 4 to any other integer would introduce a contradiction that would collapse the formal system of arithmetic.

Of course two drops of water added to two drops make one drop, but that's only because the laws of arithmetic don't apply to drops. Two plus two is always four precisely because it is empty of empirical content. It applies to cows only if you add a correspondence rule that each cow is to be identified with 1. The Pythagorean theorem also is timelessly true in all possible worlds because it follows with certainty from the symbols and rules of formal plane geometry.

Hersh's worst attack on the eternal absolute validity of arithmetic is by way of a building with no thirteenth floor. If you go up eight floors in an elevator, then five more floors, you step out of the elevator on floor 14. Hersh seems to think this makes 8 + 5 = 14 an expression that casts doubt on the validity of arithmetic addition! I might just as well cast doubt on 2 + 2 = 4 by replacing the numeral 4 with the numeral 5.

Hersh imagines that because the concept of "number" has been steadily generalized over the centuries, first to negative numbers, then to imaginary and complex numbers, quaternions, matrices, transfinite numbers, and so on, somehow makes 2 + 2 = 4 debatable. It is not debatable because it applies only to positive integers. "Dropping the insistence on certainty and indubitability," Hersh tells us, "is like moving off the [number] line into the complex plane." Baloney! Complex numbers are different entities.

Their rules have no effect on the addition of integers. Moreover, laws governing the manipulation of complex numbers are just as certain as the laws of arithmetic.

Within the formal system of Euclidean geometry, as made precise by Hilbert and others, the interior angles of a triangle add to 180 degrees. As Hersh reminds us, this was Spinoza's favorite example of an indubitable assertion. I was dumbfounded to come upon pages where Hersh brands this theorem uncertain because in non-Euclidean geometries the angles of a triangle add to more or less than a straight angle. Non-Euclidean geometries have nothing to do with Euclidean geometry. They are entirely different formal systems. Euclidean geometry says nothing about whether space-time is Euclidean or non-Euclidean. Hersh's claim of triangular uncertainty is like saying that a circle's radii are not necessarily equal because this is falsified by an ellipse.

Hersh devotes two chapters to great thinkers he believes were "humanists" (social constructivists in their philosophy of mathematics). It is a curious list. Aristotle is there because he pulled numbers and geometrical objects down from Plato's transcendent realm to make them properties of things, but to suppose he thought those forms existed only in human minds is to misread him completely. Euclid is also deemed a humanist without the slightest basis for so thinking.*

Locke is there because he recognized the obvious fact that mathematical objects are inside our brains. But Locke also believed—Hersh even quotes this!—that "the knowledge we have of mathematical truths is not only certain but

---

*The person most deserving to be on Hersh's list of maverick anti-realists is the mathematician Raymond Wilder. He and his anthropologist friend Leslie White were leading boosters of the notion that mathematical objects have no reality outside human culture. Hersh calls White's essay "The Locus of Mathematical Reality" a "beautiful statement" of social constructivism. My criticism of White's paper is reprinted in my recent anthology *The Night is Large*.

real knowledge; and not the bare empty vision of vain, insignificant chimeras of the brain." The angles of a mental triangle add to 180 degrees. This is also true, Locke adds, "of a triangle wherever it really exists." A devout theist, Locke would have been as mystified as Aristotle by the notion that mathematics has no reality outside human minds.

The inclusion of Charles Peirce as a social constructivist is even harder to defend. "I am myself a Scholastic realist of a somewhat extreme stripe," Peirce wrote (*Collected Papers*, vol. 5, p. 323). (All of Hersh's arguments against realism, by the way, were thrashed out by medieval opponents of realism.) In volume 4 (p. 89), Peirce speaks of "the Platonic world of pure forms with which mathematics is always dealing." In volume 1 (p. 350) we find this passage:

> If you enjoy the good fortune of talking with a number of mathematicians of a high order, you will find the typical pure mathematician is a sort of Platonist. . . . The eternal is for him a world, a cosmos, in which the universe of actual existence is nothing but an arbitrary locus. The end pure mathematics is pursuing is to discover the real potential world.

Hersh devotes many excellent chapters to summarizing the history of mathematics, and he ends his book with crisp, expertly worded accounts of famous mathematical proofs. The funny thing about this final chapter is that Hersh writes as if he were a realist. This is hardly surprising because the language of realism is by far the simplest, least confusing way to talk about mathematics.

Over and over again Hersh speaks of "discovering" mathematical objects that "exist." For example, the square root of 2 doesn't exist as a rational fraction, but it does exist as an irrational number that measures the length of a unit square's diagonal. Mathematicians "find" the complex numbers as "already there" on the complex plane. A set constructed by Bertrand Russell is shown not to exist. After saying that Sir William Rowan Hamilton "found" quaternions while he was crossing a bridge, Hersh reminds us that quaternions did not exist until Hamilton "discovered them." Hersh means until Hamilton "constructed them" on the basis of a social consensus of ideas, but his wording shows how easily he lapses into the language of realism.

We must constantly keep in mind that although Hersh often talks like a realist, his words have different meanings than they have for a realist. Once humans have invented a formal system like plane geometry or topology, the system can imply theorems that are difficult to "discover." Their discovery is of theorems that can be said to "exist" outside any individual mind, but with no reality beyond the collective minds of mathematicians.

Hersh closes this chapter with a beautiful new proof by George Boolos of Gödel's famous theorem that formal systems of sufficient complexity contain true statements that can't be shown true within the system. Boolos's proof is flawless, a splendid example of mathematical certainty, as are all the other proofs in this admirable chapter.

Solomon W. Golomb, writing on "Mathematics After Forty Years of the Space Age" (in *The Mathematical Intelligencer*, vol. 21, fall 1999), expressed well the Platonism of the vast majority of mathematicians past and present:

> If the Big Bang had gone slightly differently, or if we were able to spy on an entirely different universe, the laws of physics could be different from the ones we know, but 17 would still be a prime number. I recently found a very similar view attributed to the late great Julia Robinson (1919–1985) in the biography *Julia, a Life in Mathematics*, by her sister, Constance Reid. "I think that I have always had a basic liking for the natural numbers. To me they are the one real thing. We can conceive of a chemistry that is different from ours, or a biology, but we cannot conceive of a different mathematics of numbers. What is proved about numbers will be a fact in any universe.
>
> Platonism (i.e., "Realism") about mathematics has dissenters. Some who, in my view, are overly influenced

by quantum mechanics, would argue that $2^P - 1$, where P is some very large prime number, is neither prime nor composite but in some intermediate "quantum state," until it is actually tested. Of course, the Realist view is that it is already one or the other (either prime or composite), and we find out which when we test it. Even less palatable to most mathematicians is the "postmodern" criticism of all of "science," that it is just another cultural activity of humans and that its results are no more absolute or inevitable than works of poetry, music, or literature. The extreme form of this viewpoint would assert that "1 ι 7 – 11" is merely a cultural prejudice. I will readily concede the obvious: it requires a reasoning device like the human brain (or a digital computer) to perform the sequences of steps that we call "mathematics." Also, culture can play an important role in determining which mathematical questions are asked, and which mathematical topics are studied. (Our widespread use of the decimal system is undoubtedly related to humans having ten fingers.) What I will *not* concede is that, if the same mathematical questions are asked, the answers would come out, inconsistently in another culture, on another planet, in another galaxy, or even in a different universe. For example, the Greeks were interested in "perfect numbers," numbers like 6 (= 1 + 2 + 3) and 28 (= 1 + 2 + 4 + 7 + 14) which equal the sum of their exact divisors (less than the number itself). I can readily imagine a "civilization" with advanced mathematics in which the notion of "perfect numbers" was never formulated. What I cannot imagine is a civilization in which perfect numbers were defined the same way as we do, but where 28 was no longer perfect.

Let the great British mathematician G. H. Hardy have the final say:

I believe that mathematical reality lies outside us, that our function is to discover or *observe* it, and that the theorems which we prove, and which we describe grandiloquently as our "creations," are simply our notes of our observations. This view has been held, in one form or another, by many philosophers of high reputation from Plato onwards, and I shall use the language which is natural to a man who holds it. A reader who does not like the philosophy can alter the language: it will make very little difference to my conclusions.

POSTSCRIPT

As to be expected, Reuben Hersh was much annoyed by my review. He wrote a reply that was published on the Internet in October 1997. I do not know if this or some other reply has been published elsewhere.

Social constructivism, applied to science and math, continues to flourish as a kind of cult, but mainly among liberal arts professors and a few social scientists. Only a tiny number of physicial scientists, mathematicians, and philosophers of science take social constructivism seriously.

Bertrand Russell, Willard van Orman Quine, Karl Popper, and a raft of other modern philosophers are realists in the sense that they regard the physical world as "out there," and the objects and theorems of mathematics, as discovered, not invented. For a realist, supposing that our little minds create, say, the spiral structure of a galaxy, is like a child who thinks wind is caused by waving tree branches and fluttering leaves.

"It is important to realize," Popper and John Eccles write in their book *The Self and Its Brain* (1977, pp. 41–42), "that the objects and unembodied existence of these [mathematical] problems precedes their conscious discovery in the same way the existence of Mount Everest preceded its discovery."

The literature defending mathematical realism is vast. For a good defense, I recommend Penelope Maddy's *Realism in Mathematics* (1992). The book's thesis, as she capsules it, is: "Mathematics is the study of objectively existing mathematical entities just as physics is the study of physical entities. The statements of mathematics are true or false depending on the properties of those entities, independent of our ability, or lack thereof, to determine which."

For a recent defense of social constructivism in mathematics, check Paul Ernest's *Social Constructivism as a Philosophy of Mathematics* (1998). For strong criticism of this book, along with similar bashing of Hersh's book, see Bonnie Gold's review in *American Mathematical Monthly* (April 1999).

According to modern particle theory and quantum mechanics, the entire universe, at bottom, is made of mathematics. If you think otherwise, then please tell me what you think the fundamental particles and their fields *are* made of. This suggests that physics and mathematics are two ways of studying the same reality. Such a thesis is brilliantly defended by Bruno W. Augenstein, of the Rand Corporation, in his paper "Links Between Physics and Set Theory," in *Chaos, Solitons, and Fractals*, vol. 7 (1996), pp. 1761–98. The surprise so often expressed over the fact that mathematics fits so smoothly the outside world is no more surprising, Augenstein writes, than the "astonishing" fact that "cats have holes in their fur at exactly the places where their eyes are."

THE NUMBER SENSE

S tanislaw Dehaene, a neuroscientist at a Paris medical research institute, is, I regret to say, as down on mathematical realism as Reuben Hersh, whose book *What is Mathematics, Really?* I review in the preceding chapter. Dehaene's *Numbers Sense* is subtitled *How the Mind Creates Mathematics* (Oxford, 1997). However, instead of seeing numbers as social constructs, he thinks they are present at birth inside each person's head where the brain is genetically wired to recognize them. How the brain can recognize numbers unless in some sense they are outside the brain is not made clear.

Dehaene is convinced that many animals—mice, raccoons, dolphins, pigeons, parrots, and of course apes—also are capable of understanding very small integers and to perform simple arithmetic with them. His first chapter is

This review appeared in the *Los Angeles Times Book Review*, October 12, 1997.

a fascinating survey of recent experiments with animals that seem to support this view. Rats, for example, can be trained to take the fourth entrance to any maze. They can be trained to press a lever exactly n times if n is a very small number.

In one experiment a rat was conditioned to press a certain lever only after it heard two tones followed by two flashes of light. Dehaene takes this as evidence that a rat somehow knows that $2 + 2 = 4$. A chimp named Sheba, after finding the numeral 2 under a table, and the numeral 4 inside a box, was trained to select the numeral 6 from a set of cards. To Sarah Boysen, who designed this and similar sensational experiments, it shows that chimps can add 2 and 4 to get 6. If true, it proves that mathematics is not restricted to *human* cultures.

"Never had an animal come any closer," Dehaene writes, "to the symbolic calculation abilities exhibited by humankind." I'm inclined to doubt this. Assuming Sheba was not responding to unconscious cues from her trainers, it seems to me possible she was simply associating a combination of two meaningless symbols with a third symbol without the slightest inkling of their arithmetic meanings.

In his next chapter, "Babies Who Count," Dehaene describes experiments with babies which he is convinced show that they can do simple arithmetic long before they can speak. Most of this research fails to impress me. For instance, four-and-a-half-month-old infants are shown two Mickey Mouse toys. A screen is placed to conceal them. Behind the screen, two red balls are substituted for the Mickey Mouse toys. When the screen is removed, and the babies see the two balls, they are not in the least surprised. But if they see only one red ball or three red balls, they appear "shocked."

"Mickey Mouse turning into a ball . . . ," Dehaene writes, "is an acceptable transformation as far as the baby's number processing system is concerned. As long as no object vanishes or is created *de novo*, the operation is judged to be numerically correct and yields no surprise reaction in babies." Twoness is preserved. But if two objects turn magically into one or three, the baby is startled. "The demonstration is irrefutable," the author says. "Babies know that 1 + 1 makes neither 1 nor 3, but exactly 2."

How did psychologist Tony Simon, who supervised this test, decide when a baby is surprised? By measuring the average amount of time a baby takes when looking at objects. If it takes a few seconds longer when it sees that two objects have changed to one or three, this is taken to indicate the baby is "shocked." In one experiment with Mickey Mouse the average time a baby looked at the mouse, before a change occurred, was thirteen seconds. After the change, it took half a second longer! Now, measuring and averaging such times is not easy because babies look here and there, seldom keeping a fixed gaze on anything. There is so much room here for an experimenter's expectations to bias statistics that I find it hard to accept this and similar tests as proof that young babies, like chimps, have an inherited ability to do simple arithmetic.

Jean Piaget's famous experiments with infants showed that children have to be several years old before they have any grasp of numbers. Dehaene believes Piaget was wrong. "Babies are much better mathematicians than we thought only fifteen years ago," he assures us. "When they blow the first candle on their birthday cake, parents have every reason to be proud of them, for they have already acquired whether by learning or by mere cerebral maturation, the rudiments of arithmetic and a surprisingly articulate 'number sense.'"

The rest of Dehaene's book concerns the history of numbers, variations in the counting practices of different cultures, the mystery of idiot-savants who perform rapid calculations with huge numbers, the intricate structure of the human brain, and how its workings differ from those of digital computers. Not until his final chapter, "What is a Number?" does his antirealism become explicit.

Platonic realism is branded a philosophy "no neuro-biologist can believe." Most mathematicians, he grants, "feel" as if they are exploring a jungle "out there," but this is sheer illusion. Formalism is dismissed because it turns mathematics into a game played with strings of meaning-less symbols. He forgets that those symbols can be inter-preted as mathematical objects which in turn can be applied to the outside world. His sympathies are with the philosophical school called intuitionism (now called con-structivism) which views mathematical objects as "nothing but constructions of the human mind."

Nothing but? Dehaene now bumps into what he calls the "unfathomable mystery" of why mathematics tells us so much about the universe. "How is it possible," Einstein is quoted, "that mathematics, a product of human thought that is independent of experience, fits so excellently the objects of physical reality?" Dehaene does not add how Einstein answered. The Old One, as he liked to call the uni-verse, is mathematically structured.

Dehaene admits that "reality is organized according to structures that predate the human mind," but he cannot bring himself to say, as everybody else would say, that those structures are mathematical. If we restrict mathematical objects and theorems to what mathematicians say and write, it is trivially true that mathematics is entirely a human endeavor. We might just as well say that the laws of physics are limited to what physicists say and write, wtih no locus outside human culture. It's hard to believe, but a few sociol-ogists actually think that physical laws are entirely human inventions! The nontrivial question is whether it is mean-ingful to assert (I choose from millions of examples) that the shell of a dinosaur egg divided space into inside and outside long before a topologist evolved to prove it.

Like Hersh and other antirealists, Dehaene loves to stress the uncertainty of mathematics. To demonstrate this he closes his book by exhuming one of the oldest fal-lacies in geometry, a seeming proof that a right angle equals an obtuse angle. Euclid wrote a book about such fallacies—alas, it did not survive—which may well have included this amusing chestnut. I first ran across it in high school in a book on mathematical recreations. Does it suggest that plane geometry is uncertain? No, it shows exactly the opposite. Once the faulty construction of the diagram is recognized, it follows with iron certainty that the two angles are *not* equal.

Is the universe written in mathematical language, as Galileo said? Dehaene thinks not. All mathematical objects "are mental constructions whose roots are to be found in the adaptation of the human brain to the regularities of the universe." True enough, but why refuse to say, as almost all mathematicians and everyone else says, that those regularities are mathematical?

It is not just that planets tend to move in elliptical orbits or that galaxies often spiral, but matter itself has now dissolved into pure mathematics. The entire universe, including you and me, is made of leptons and quarks, particles that are the quantized aspects of fields. Perhaps all particles are made up of still smaller particles called superstrings. And what are the fields and particles made of? Patterns. Nothing can be said about these patterns except to describe their mathematical properties. You and I, like the stars, are made of mathematics. "The universe seems to be made of nothing," a friend recently remarked, "yet somehow it manages to exist." This is what Ronald Graham, of Bell Labs, meant when he once declared, "Mathematics is the *only* reality."

You would think that the testimonies of so many eminent mathematicians about the independence of mathematical objects from human culture might arouse a bit of humility, wonder, and mystery among the antirealists who make humanity the center of all being. Although it will have not the slightest effect, let me close with some statements made last year by the great John Conway, now at Princeton, in a radio interview for the Canadian Broadcasting System:

I don't think there's much disagreement about mathematics. We discover it. We might invent a particular method for solving a problem. So *some* mathematics is invented, but the real cornerstones of mathematics are discovered. They're out there. . . . Take the whole numbers, 1, 2, 3. People often call those "concepts" . . . implying that they're just in the mind. But they're not! And so if we study those numbers we are studying an abstract part of the world, rather than some of its more concrete aspects. But anybody who studies numbers is studying something that is really there.

Of course they are not "there" in the same way the Moon is there. But the patterns of which the Moon is made are eternal, more real than the poor old Moon itself destined like you and me to eventually vanish from the universe.

POSTSCRIPT

Reuben Hersh reviewed Dehaene's book in *American Mathematical Monthly* (vol. 105, December 1998, pp. 975–76). The review, of course is favorable. "After reading it," Hersh writes, "I am convinced that something in our brains underlies and makes possible mathematical thinking. Brain researchers are beginning to find out what it is and where it is."

Hersh quotes the following passage from Dehaene's book which expresses his own antirealist view:

For an epistemologist, a neurobiologist, or a neuropsychologist, the Platonist position seems hard to defend. . . . Even if mathematicians' introspection convinces them of the tangible reality of the objects they study, this feeling cannot be more than an illusion. Presumably, one can become a mathematical genius only if one has an outstanding capacity for forming vivid mental representations of abstract mathematical concepts—mental images that soon turn into an illusion, eclipsing the human origins of mathematical objects and endowing them with the semblance of an independent existence.

One can understand why a neuroscientist, whose specialty is studying the human brain, might come to think that all of mathematics is inside that brain, but that a mathematician as distinguished as Reuben Hersh would agree, is not so easy to understand.

INTRODUCTIONS TO THREE BOOKS BY H. G. WELLS

THE CONQUEST OF TIME*

In 1908, when H. G. Wells was forty-two, he published *First and Last Things*, a book in which he defended for the first time his central philosophical beliefs. The universe is out there, independent of us. Our minds, however, are "clumsy forceps" that can only partially grasp truth. Science is fallible. Every word is fuzzy: "In cooperation with an intelligent joiner I would undertake to defeat any definition of chair or chairishness that you gave me."

There are truths beyond our comprehension, Wells wrote, such as a resolution of the apparent contradiction between free will and determinism. Behind the world we know there seems to be an order, a "scheme," even though we cannot know what it is.

---

*This introduction first appeared in H. G. Wells, *The Conquest of Time* (New York: Dover, 1995).

Wells could not believe he was important enough in this vast scheme of things to survive death. Our task is not to earn a place in heaven, but to do what we can to eliminate war and forge a world of peace, justice, and beauty. Wells calls this socialism. To make the task endurable we must not take it too seriously. "Behind everything I perceived the smile. In the ultimate I know, though I cannot prove . . . that everything is right and all things mine."

In 1941, when Wells was seventy-five, he wrote a second confessional, *The Conquest of Time*. As he said on the title page of the first edition, it was a book to replace *First and Last Things*. That earlier book had been written, Wells says in his opening chapter, when he was "still mentally adolescent." It "threw out sterile flowers, bright ideas that led nowhere."

None of the bright ideas in *First and Last Things* is abandoned in *The Conquest of Time*; but the latter is a better written, more compact, clearer expression of Wells's fundamental beliefs. Conquering time becomes a metaphor for humanity's steady overcoming of the ills, tyrannies, and superstitions of the past as the world moves forward toward a future in which war will be no more and all humanity united in a common brotherhood.

First and Last Things expressed a hope that Christianity would move steadily in the direction of secular humanism. In *The Conquest of Time* Wells discards this tolerance for a vigorous attack on Christianity, especially in its Roman Catholic form. A five-page footnote on Jesus (pp. 34–38) is a remarkably accurate summary of what modern biblical scholars have to say about the historical Jesus and the contrast between what he taught and the harsh, bizarre doctrines of St. Paul. Now in his old age, Wells does not hesitate to say openly that the "Christian God of hell was an utterly detestable maniac."

The themes of *First and Last Things* reappear here in more sophisticated ways. Although science is never absolutely certain, induction enables us to get closer and closer to final truth. The universe may be indifferent to our fate, "but it seems to play fair upon some vaster system of its own. It honors many of our inferences; it confirms our reasoned prophesies" (p. 67).

Wells sees humanity as evolving slowly toward a new species he calls After-Man, whose religion will be secular humanism. Wells's earlier confessional did not object to using the word "God" provided it did not imply anything resembling human personality. In his middle years Wells went through a phase of defending the concept of a "finite God," similar to the god of such process theologians as Samuel Alexander, Edgar Brightman, and Charles Hartshorne. (He even wrote a book about it called *God the Invisible King*.) Later Wells decided he was really an atheist. He would have agreed with Sidney Hook's criticism of John Dewey for his Pickwickian use of the term "God" in *A Common Faith*. Future humanity, Wells writes here, "cannot afford to recognize any prevaricating use of the word 'God.' That word implies a personality or it implies nothing" (p. 77).

The new "religion," for Wells, assumes that the "whole object of life is the progressive conquest of hunger and thirst, of climate, of substance, of mechanical power, of bodily and mental pain, of space and distance, of time, of things that have seemed lost in the past and of things possible in the future."

Wells closes his little book with speculations drawn from the theory of relativity. He embraces the conviction, which Einstein shared with Spinoza, that both past and future exist permanently in space-time. Einstein's four-dimensional universe "is rigid, Calvinistic, predestinate" (p. 99).

Wells sees no conflict between fate and free will. "The major aspect of life is Destiny; the minor is that we do not know our destiny" (p. 99). To put it plainly, although we must act as if our wills were free, in a deeper sense this is an illusion: "Our conscious lives are like the pictures on a cinema screen." Thus Wells, in his final years, accepted what William James derisively called a "block universe," frozen forever in space-time like a monstrous block of wood. Apparently Wells was unfamiliar with how quantum mechanics had replaced determinism with genuine randomness that permeates the universe on its microlevel.

Wells lived to witness the fall of atom bombs, an event he had amazingly predicted as early as 1914 in his science fiction novel *The World Set Free*. He even used the term "atom bomb" when describing the discovery of atomic energy and its military use in a world war taking place in the 1940s! Throughout his life Wells alternated between optimism and pessimism, but optimism was always his dominant mood. The splitting of the atom, however, plunged him into despair.

In *Mind at the End of Its Tether* (1946) Wells writes of a "frightful queerness" that has come over the world, a "new cold glare mocks and dazzles human intelligence. . . . The writer is convinced that there is no way out or around or through the impasse. It is the end." Humanity is about to self destruct. His book, Wells says, brings to an end everything he has written: "He has nothing more and never will have anything more to say."

Knowing he was dying, Wells projected onto humanity his own impending doom. The book you now hold was written by a healthier man. It remains the best record of what a mature Wells truly believed about life and death and the universe.

THE COUNTRY OF THE BLIND AND OTHER SCIENCE FICTION STORIES*

There is no need to sketch here the life of an author as well known as H. G. Wells (1866–1946). If you care to learn the biographical details, you can find them in encyclopedias,

---

*This introduction first appeared in H. G. Wells, *The Country of the Blind and Other Science Fiction Stories* (New York: Dover, 1997).

in books about modern British authors, in more than a dozen biographies and in Wells's 1934 autobiography, *Experiment in Autobiography: Discoveries and Conclusions of a Very Ordinary Brain (Since 1866)*.

As everyone knows, Wells was a prolific writer of novels, short stories, and nonfiction. The most complete collection of his short fiction, *The Short Stories of H. G. Wells*, was published in London in 1927. There have since been many less complete anthologies. *The Man with a Nose* (1984), edited by J. R. Hammond, contains nineteen short stories that do not appear in any previous collection.

The book you now hold is a selection of six of the most famous short tales by Wells. I have not included "The Time Machine" and "The Invisible Man" because they are novellas (available in Dover Thrift Editions: 28472-7 and 27071-8, respectively). I have also ignored such well-known stories as "The Man Who Could Work Miracles" because they are closer to fantasy than to science fiction.

I have written an introduction to each story to document its first publication and to make relevant comments. It is interesting to note that among Wells's many realistic stories, almost none is worth remembering.

The Country of the Blind

Wells's original version of "The Country of the Blind" was first published in *The Strand Magazine* (May 1904), with eight illustrations. At the story's end Nunez, unwilling to have his eyes removed, tries to climb out of the South American valley into which he had tumbled. "He thought of Medina-saroté, and she had become small and remote." Here is the tale's tragic conclusion:

> When sunset came he was no longer climbing, but he was far and high. His clothes were torn, his limbs were blood-stained, he was bruised in many places, but he lay as if he were at his ease, and there was a smile on his face.
>
> From where he rested the valley seemed as if it were

in a pit and nearly a mile below. Already it was dim with haze and shadow, though the mountain summits around him were things of light and fire. The mountain summits around him were things of light and fire, and the little details of the rocks near at hand were drenched with subtle beauty—a vein of green mineral piercing the grey, the flash of crystal faces here and there, a minute, minutely beautiful orange lichen close beside his face. There were deep, mysterious shadows in the gorge, blue deepening into purple, and purple into a luminous darkness, and overhead was the illimitable vastness of the sky. But he heeded these things no longer, but lay quite inactive there, smiling as if he were satisfied merely to have escaped from the valley of the Blind in which he had thought to be King.

The glow of the sunset passed, and the night came, and still he lay peacefully contented, under the cold stars.

In 1939 Wells revised his story for a limited edition of 280 copies published by London's Golden Cockerel Press. Although the revised version is almost a short novel— Wells added some three thousand words—I give it here because very few Wells admirers are even aware of its existence. In this book's introduction Wells wrote:

> The stress [in the original version] is upon the spiritual isolation of those who see more keenly than their fellows and the tragedy of their incommunicable appreciation of life. The visionary dies, a worthless outcast, finding no other escape from his gift but death, and the blind world goes on, invincibly self-satisfied and secure. But in the later story vision becomes something altogether more tragic; it is no longer a story of disregarded loveliness and release; the visionary sees destruction sweeping down upon the whole blind world he has come to endure and even to love; he sees it plain, and he can do nothing to save it from its fate.

Although both stories are beautifully written and emotionally moving, it is their metaphorical level that elevates them to greatness. Wells must have had in mind Plato's famous allegory of the cave in the seventh book of *The Republic*. Socrates asks Glaucon to imagine the fate of a man who goes outside the cave—a cave where one can see only shadows on a wall—into the great outside world of sunlight. Imagine, Socrates says, that man returning to the cave's darkness:

> And if there were a contest, and he had to compete in measuring the shadows with the prisoners who had never moved out of the den, while his sight was still weak, and before his eyes had become steady (and the time which would be needed to acquire this new habit of sight might be very considerable), would he not be ridiculous? Men would say of him that up he went and down he came without his eyes; and that it was better not even to think of ascending; and if any one tried to loose another and lead him up to the light, let them only catch the offender, and they would put him to death.
>
> No question, he said.
>
> This entire allegory, I said, you may now append, dear Glaucon, to the previous argument; the prison-house is the world of sight, the light of the fire is the sun, and you will not misapprehend me if you interpret the journey upwards to be the ascent of the soul into the intellectual world according to my poor belief, which, at your desire, I have expressed—whether rightly or wrongly God knows. But, whether true or false, my opinion is that in the world of knowledge the idea of good appears last of all, and is seen only with an effort; and, when seen, is also inferred to be the universal author of all things beautiful and right, parent of light and of the lord of light in this visible world, and the immediate source of reason and truth in the intellectual; and that this is the power upon which he who would act rationally either in public or private life must have his eye fixed.
>
> I agree, he said, as far as I am able to understand you.
>
> Moreover, I said, you must not wonder that those who attain to this beatific vision are unwilling to descend to human affairs; for their souls are ever hastening into the upper world where they desire to dwell; which desire of theirs is very natural, if our allegory may be trusted.
>
> Yes, very natural.

And is there anything surprising in one who passes from divine contemplations to the evil state of man, misbehaving himself in a ridiculous manner; if, while his eyes are blinking and before he has become accustomed to the surrounding darkness, he is compelled to fight in courts of law, or in other places, about the images or the shadows of images of justice, and is endeavoring to meet the conceptions of those who have never yet seen absolute justice?

Anything but surprising, he replied.

Any one who has common sense will remember that the bewilderments of the eyes are of two kinds, and arise from two causes, either from coming out of the light or from going into the light, which is true of the mind's eye, quite as much as of the bodily eye; and he who remembers this when he sees any one whose vision is perplexed and weak, will not be too ready to laugh; he will first ask whether that soul of man has come out of the brighter life, and is unable to see because unaccustomed to the dark, or having turned from darkness to the day is dazzled by excess of light. And he will count the one happy in his condition and state of being, and he will pity the other; or, if he have a mind to laugh at the soul which comes from below into the light, there will be more reason in this than in the laugh which greets him who returns from above out of the light into the den.

The allegory, as Wells presents it, is a powerful one that applies to all true believers in a fixed ideology, whether political or religious, who are blind to knowledge and reason. Was not the Soviet Union, in the grip of Karl Marx's crank economics, for decades a country of the blind? Were not Soviet leaders of vision executed or banished to Siberia by the blind and cruel Stalin? In post-Stalin Russia hundreds were even confined to mental hospitals in the belief that anyone who opposed the official ideology must be mentally unbalanced!

Was not Hitler's Germany another country of the blind? Is not today's China, still in the grip of obsolete Marxism, another blind country? As in Stalin's Russia, those who see clearly the evils of Chinese totalitarianism are still being killed or imprisoned by blind leaders.

Wells's allegorical tale applies with equal force to dogmatic religions and cults. Extremist Catholics, Jews, Protestants, and Moslems, and the unshakable members of a hundred religious cults around the world, do not know they are blind. It is almost impossible to persuade them by argument that there is a wider world of light. "You cannot even fight happily," Nunez reflects, "with creatures that stand upon a different mental basis to yourself."

So persuasive and strong is the rhetoric of the blind that even men and women of vision are often tempted to believe. One thinks of that terrible scene in George Orwell's *Nineteen Eighty-Four* in which Winston, after hours of brainwashing, actually imagines for a fleeting instant that when O'Brien holds up four fingers he sees five—a proof by O'Brien that even mathematical truth is what the Party proclaims. Two plus 2 can equal 5 or 3 depending on what the Party finds expedient.

Nunez, passionately in love with Medina-saroté, comes close to accepting surgery that will deprive him of sight. The blind residents of the valley believe that their world is covered by a stone roof above which is the god they call the Wisdom Above. So impressive are their arguments that for a moment, like Orwell's Winston, Nunez "almost doubted whether indeed he was not the victim of hallucination in not seeing it [the lid of rock] overhead."

We all know persons who, because they are living in a region of the blind, or are married to a resident of such a region, find it necessary to adapt, to pretend to believe in the ideology of a loved spouse or of friends around them. In Wells's first version of his story, Nunez prefers death by exposure to the loss of his eyes. In the revised version, after discovering a mammoth rockslide that will destroy the Country of the Blind, he manages to escape with Medina-saroté before the doom falls.

Although Medina-saroté adjusts well enough to Nunez's world to raise his children, she never fully escapes from the mindset of her past. She does not want to lose her faith in the Wisdom Above. The outside world may be beautiful, she admits, in the story's poignant ending, but she refuses medical help that might give her vision. She still thinks it must be terrible to see. "If I am dreaming," so goes the old evangelical hymn, "let me dream on."

The Star

Many science-fiction novels and stories concern catastrophes that would befall the earth if it were ever struck, or narrowly missed, by a giant comet, asteroid, or some other huge celestial object. Wells himself devoted an entire novel, *In the Days of the Comet*, to such an event. However, no short story has ever described what might happen with such riveting vividness as Wells's "The Star."

Strictly speaking, the object is not a star but a dark sister planet beyond the orbit of Neptune. (Pluto had not been discovered when Wells wrote.) For some reason the planet has been shifted from its orbit and is headed toward the sun. On its way it collides with Neptune. The two coalesce to form an incandescent "star" that passes extremely close to the earth before it strikes the sun.

The story is timely. In recent years there has been much speculation about the possibility of a large asteroid hitting the earth—there have been several fairly close encounters—and what steps could be taken to avert the earth's destruction.

In his last paragraph Wells introduces Martian astronomers who watch the star's near collision with our planet. Were he alive today, Wells would probably put his astronomers on Titan. This giant moon of Saturn—it is larger than Mercury—is the only satellite in the solar system with an atmosphere that just might harbor, but probably does not, intelligent life.

"The Star" first appeared in the 1897 Christmas

number of London's *The Graphic*. A full-page, full-color illustration showed a street packed with Londoners staring at the sky and shouting "it is brighter!" A newsboy holds a sheet with huge red letters that say "Total Destruction of the Earth."

The New Accelerator

I can still remember the excitement of first reading this story when I was twelve. As a charter subscriber to Hugo Gernsback's *Amazing Stories*, I read "The New Accelerator" when it was reprinted in the first issue (April 1926) of the first magazine devoted exclusively to science fiction. Would that I had preserved those early issues! They would be worth a lot of money to collectors today.

That one could increase the velocity of electrical impulses through the nervous system and the brain may not be too outlandish a notion, but a heart beating a thousand times faster presents serious difficulties. Blood pressure would burst arteries, not to mention the heat produced by the friction of fast-flowing blood. Vocal cords would be vibrating a thousand times faster, raising the frequency of sound waves far too high for the ear to hear. Moreover, sound would still travel at the same rate through air, making it impossible for two men to have long conversations in just a few microseconds.

Of course Wells, a trained biologist, would have been aware of such problems, but facts often have to be neglected in science fiction for the sake of a good yarn. Wells himself once described his science fiction as "impossible" in contrast to the work of Jules Verne.

"The New Accelerator" first ran in *The Strand* (December 1901). This was before moving pictures accustomed audiences to slow- and fast-motion effects produced by running the camera faster or slower than usual. It was also four years before Einstein's first paper on special relativity made clear that time was relative.

In Einstein's famous clock paradox, a person could travel long distances at high speed, and only a few weeks would go by on the spaceship while years flashed by on earth. Similarly, the accelerator drug allows the narrator to write his entire story in one sitting while six minutes pass in the outside world. I suspect that my boyhood wonder at Wells's demonstration of time's relativity had an influence on my later efforts to understand the essentials of relativity theory.

The Remarkable Case of Davidson's Eyes

In my opinion, the evidence for remote viewing, or what used to he called clairvoyance, is close to zero. It certainly is true that parapsychologists have not the foggiest idea of how it could work, especially in view of their belief that it is independent of both time and distance. The explanation put forth in this early story by Wells is as good as any.

Davidson's head is between the poles of a powerful electromagnet when a lightning bolt strikes the laboratory. As a result his eyes become attached to a "kink," or what later science fiction calls a "warp," in space-time. For three weeks Davidson's eyes seem to be located on an island eight thousand miles away.

Just as two widely separated points in Flatland can be brought together by folding the plane, Wells explains, so two widely separated points in space can be brought together by folding space through a fourth dimension. Today's farout cosmologists speak seriously of "wormholes" joining parts of space by way of a higher dimension. In superstring theory, for the first time in the history of physics, higher space dimensions are considered by many physicists to be as real as three-dimensional space.

Wells was fascinated by the concept of higher dimensions. In "The Plattner Story" an explosion propels a man into the fourth dimension. When another explosion sends him back he is left-right reversed, like lifting a resident of

Flatland off the plane, turning him over and putting him down again. Wells's great Utopia novel *Men like Gods* made an early use of parallel universes lying side by side in the fourth dimension like the pages of a book.

"The Remarkable Case of Davidson's Eyes" was first published in *The Pall Mall Budget* (March 28, 1895).

Under the Knife

Asleep under the influence of chloroform while undergoing a dangerous operation, the narrator has a vivid out-of-body dream. In his dream the operation is a failure and he dies. His spirit escapes from his body, to travel outward through the solar system, through the Milky Way galaxy, and finally leaves the universe altogether. Still soaring through the blackness, he sees the universe shrink to a glittering point of light reflected from a ring on the finger of a gigantic clenched hand!

This notion that our stars are atoms in a vaster universe, "and those again of another, and so on through an endless progression," later became a common science-fiction theme. In recent times Rudy Rucker has given it a bizarre topological twist. In one of his novels he imagines that the hierarchy of universes closes on itself. As you move outward through larger worlds, or downward through smaller ones, you eventually arrive back in the universe we know.

As Wells's dreamer travels through the solar system his time slows to a point where he sees the moon revolving rapidly around the earth. The story's description of the vast reaches of space-time, and the infinitesimal speck of matter on which we live, is Wells writing at his best. The story first appeared in *The New Review* (January 1896).

The Queer Story of Brownlow's Newspaper

H. G. Wells acquired a reputation as a prophet, and deservedly so because so many of his predictions came true.

His greatest hit was his account of the discovery of how to unleash atomic energy and the use of what he called "atomic bombs" in a second world war triggered by Germany's invasion of France. All this in a novel, *The World Set Free*, published in 1914! There are many other remarkable hits in Wells's science fiction. For example, the steel sphere in his 1896 story "In the Abyss" allows a scientist to explore the bottom of the sea.

On the other hand, Wells had enormous misses. His greatest error was taking seriously the dubious reports of canals on Mars. They led Wells to introduce bizarre Martians in such novels as *The War of the Worlds* and in such stories as "The Crystal Egg." Of course hundreds of other science-fiction writers also wrote about Martians. Even in his serious nonfiction Wells made numerous bad guesses. Writing in 1901, in *Anticipations*, he did not think it probable that airplanes "will ever come into play as a serious modification of transport and communication."

In that same volume, in a chapter on twentieth-century warfare, Wells had this to say about submarines: "I must confess that my imagination, in spite even of spurring, refuses to see any sort of submarine doing anything but suffocate its crew and founder at sea." The chapter is filled with other blunders, although Wells correctly predicts that "the nation that will be the most powerful in warfare as in peace, will certainly be the ascendent or dominant nation before the year 2000."

In *The Way the World Is Going* (1928) Wells anticipated the "complete disappearance" of radio broadcasting, "confident that the unfortunate people, who must now subdue themselves to 'listening-in,' will soon find a better pastime for their leisure."

"The Queer Story of Brownlow's Newspaper" was a more mature Wells's attempt to describe England forty years in the future. It first appeared in, of all places, America's *Ladies' Home Journal* (April 1932). Almost forty years later, on November 10, 1971, London's *Evening Stan-*

dard reprinted the tale with comments on its hits and misses. Not surprisingly, no one named Evan O'Hara was found living in London. In the United States, on November 12, 1971, on the Long John Nebel radio talk show, science-fiction writers Frederik Pohl and Isaac Asimov discussed Wells's story.

Brownlow's newspaper is the usual Wells blend of good and bad prophecy. His hits are the widespread use of color in newspapers, the greater emphasis in the print media on science news and on psychological motivations, and the reduction of body clothing, though not to the extent of bare chests and breasts, or moving pockets to sleeves. Many people in Southwest Asia today do wear Western clothes. The Soviet empire has, happily, collapsed, as Wells expected, though this had not occurred by 1971.

Unfortunately, Wells's misses are far wider. Germany, France, England, and the United States still flourish as nations. There is not the slightest sign of decreasing nationalism or the rise of a world government with police powerful enough to stop crime. The age of combustion has not been replaced by heat from the earth. Spelling has not been simplified. The stock market and financial pages have not become obsolete. The gorilla is not yet extinct. No thirteen-month calendar has been adopted.

Although some nations have curbed their birthrate, the world's rate is nowhere close to Wells's optimistic guesses. Newspapers are not printed on paper that is mostly aluminum. It is hard to comprehend how Wells failed to predict television, especially in view of the fact that Hugo Gernsback, who was enthusiastically reprinting Wells in *Amazing Stories*, actually ran a TV broadcast station in the twenties. The picture was tiny and people had to build their own sets, but the technology was rapidly advancing.

Brownlow's newspaper has nothing to say about atomic energy in spite of Wells's amazing anticipation of it in his *The World Set Free*. Space flight is also absent. Wells had earlier written a novel about a voyage to the moon but considered this one of his "impossible" fictions, never dreaming that such a trip would occur before the end of the century. Our astronauts walked on the moon in 1969. Ironically, Wells was a better prophet in his science fiction than in his more careful nonfiction efforts.

In 1933 Wells's most ambitious attempt to foresee the future was his thick book *The Shape of Things to Come*. It again is a mix of good and bad prophecy. The book makes no mention of the computer revolution, space flight, the discovery of DNA, the rise of television or the atomic age.

In Wells's introduction, Philip Raven, whose "dream book" is the basis for *The Shape of Things to Come*, describes his prophetic work as "rather like the newspaper of your friend Brownlow." Herbert George Wells could have ended his book the way he ended the story about Brownlow, by saying that he is as convinced of its accuracy as he is convinced that his name is Hubert G. Wells.

*ANTICIPATIONS**

Herbert George Wells (1866–1946) was thirty-five in 1901 when this book was first published in Leipzig by Bernhard Tauchnitz. Titled *Anticipations of the Reaction of Mechanical and Scientific Progress Upon Human Life and Thought*, it had earlier been serialized in *The Fortnightly Review* (April through December 1901) with the subtitle: *An Experiment in Prophecy*. Wells was already famous for his science-fiction novels and short stories. A second edition of *Anticipations* in 1902 contained numerous revisions and additions. In 1914 Wells added a new introduction, here reprinted, to a cheaper edition of the book.

Anticipations was Wells's first best-seller. The book had an enormous impact on British intellectuals and their European counterparts. George Bernard Shaw, Sidney and

---

*This introduction first appeared in H. G. Wells, *Anticipations* (New York: Dover, 1999).

Beatrice Webb, William and Henry James, and Arnold Bennett were among a raft of eminent writers who highly praised the book.

Throughout his life Wells fancied himself a shrewd prophet of events to come. Many of his later books and articles, such as *What Is Coming?* (1914), *A Year of Prophesying* (1924), *The Way the World Is Going* (1928), *The Shape of Things to Come* (1933), and *The Outlook for Homo Sapiens* (1942) were similar to *Anticipations* in their efforts to predict the future.

Wells also explored the future in his science fiction. His wildest misses were novels and tales about Martians, and a novel about intelligent humanoids living in caves below the surface of the Moon. Several utopian fantasies, and one negative utopian novel (a nightmare vision of a possible future) are mixtures of hits and misses. It was in his 1914 novel *The World Set Free* that Wells made his most astounding hit. Dedicated to Frederick Soddy for his pioneer research on radium, the novel opens with a moving extract from the diary of a physicist who has found a way to split the atom and release atomic energy. He is fearful of the consequences of his discovery, but realizes that, had he not made it, other physicists soon would. The novel describes a war between England and Germany, in the middle of the twentieth century, during which "atomic bombs," as Wells actually called them, were dropped from airplanes.

Wells was always good at seeing the immediate consequences of new technologies, but science has a wonderful way of springing great surprises. On this account, we should not fault Wells for never anticipating, in this or any other book, the rapid development of motion pictures and television, the computer revolution, the DNA revolution, how soon astronauts would land on the Moon, or how soon space probes would be exploring our solar system.

In the first chapter of *Anticipations* Wells scored a number of hits. He predicted that coal and steam, as power sources for locomotives and ships, would soon be replaced by what he called "explosion engines" driven by gas and oil. He suggested that the piston might some day be replaced by a rotary device such as the one now used in Mazdas. He expected electrical cars to become practical. He foresaw the coming of huge trucks in competition with railroads, and the replacement of horse-drawn vehicles by cars and omnibuses.

The rapid proliferation of trucks and cars, Wells realized, would require new and wider roads, the dirt replaced by asphalt or some other hard substance. The roads would be shaped to drain rainfall, and have guardrails to prevent cars from plunging off embankments. He predicted bridges and underpasses where highways intersected, and the need for strong traffic regulations.

Train rides, Wells foresaw, would soon be smooth enough to allow dining. His suggestion that the century would see moving sidewalks, running side by side at different speeds, has failed to materialize except in the case of some large airline terminals. Wells's greatest miss in this chapter is his failure to anticipate how rapidly air travel would arrive. "I do not think it at all probable," he says in a footnote, "that aeronautics will ever come into play as a serious modification of transport and communication. . . . Man is not, for example, an albatross, but a land biped. . . ."

Wells next turns his attention to the earth's monstrous cities. Borrowing terms from physics, he considers the centripetal forces causing congestion, and the centrifugal forces sending city dwellers to the suburbs. He rightly guesses that trains and cars will intensify this dispersion, with telephones also playing a role. Unfortunately, Wells got carried away with his estimation of the dominance of centrifugal forces. The great cities, he predicted, "are destined to such a process of dissection and diffusion as to amount almost to obliteration. By the year 2000, he writes, it is probable that London's "urban region" would include almost all of England and parts of Wales. In Eastern United States, he predicted that the "vast stretch of country" from

Washington, D.C., to Albany, would become home for citizens working in New York City and Philadelphia.

Wells hit the mark in sensing that increasing numbers of business firms would move from cities to uncrowded towns and suburbs. Alas, what he calls the "pauper masses" still huddle in most of the world's metropolises; the thousands of sick and homeless creating social problems as seemingly unsolvable today as they have been in the past.

Chapter 3, on changing social conditions, predicts the rise in advanced nations of several classes new to history: There will be stockholders who "do nothing in common except receive and hope for dividends." There will also be a class made up of the working poor, whom Wells calls "a multitude of people drifting down towards the abyss," victims of new technologies for which they are untrained. Wells sees this group being replaced by machines or thrown out of work by the flight of companies to lands where labor is cheaper. Finally, Wells expects a vast, chaotic, educated middle class of professional specialists.

Wells realized that the rapid progress of science would alter the social fabric of advanced cultures in unpredictable ways. Chapter 3's chief value is its vivid portrayal of British life at the time of its writing. Two gems are an hilarious footnote describing a meeting of the House of Commons, and Wells's account of the bumbling process by which one of his houses was built.

It is hard to fault most of the predictions in chapter 4. Wells describes a typical middle-class home of the future as one surrounded by a yard, and centrally heated by warm air blown from wall ducts. Oil lamps, he expects, will be replaced by electric lights. Stoves will be electric. Chimneys will either vanish or remain rising from bogus fireplaces with fake glowing logs.*

---

*Parade (November 15, 1998) ran a half-page ad for "The Ultimate Fireplace Video." For six hours the TV screen shows a picture of burning, crackling logs. "It's soothing," reads the ad. "It's romantic. And it's maintenance free. . . . No chopping wood, no fancy equipment to buy. No cinders and soot to clean. And no smoky living room." And, of course, no heat.

(Wells stumbled in anticipating fake smoke rising from the sham chimneys.) Every bedroom will have an adjoining bathroom. Men will no longer feel obliged to wear boots indoors.

Rich shareholders—Wells calls them the "leisure class"—will control architecture, art, and fashion. Striking new styles of clothing will come and go. Popular novels and plays will reflect the doings of the leisure class. There will be nursery schools for the very young.

Moral restrictions will decline as men and women seek greater sexual freedom. Divorces will increase. There will be more childless marriages, more children born out of wedlock. Regions of "opulent enjoyment," like the French Riviera resorts of Wells's day, will flourish. The loss of moral and religious certainty will create great confusion and unrest until stable patterns of ethics emerge unencumbered by stale religious creeds.

Two whimsical misses are worth noting in this chapter. Wells suggests that windows of the future will be cleaned by jets of soapy water flowing from holes in a horizontal pipe above each window. To ease the cleaning of floors he expects sharp angles between floors and walls to be covered by curved surfaces.

Wells's fifth chapter contains the first of many attacks he was to make on what he sees as degenerate forms of democracy responsible for poor government and terrible wars. Wells had no respect for a system in which power is in the hands of ignorant voters easily swayed by demagogs. There is, he insists, "nothing in the mind of the average man except blank indifference." For the rest of his life Wells hoped and believed that mass democracy would be replaced by a socialist government in the hands of an aristocracy of well-educated, scientifically minded men. They would take power, Wells believed, like butterflies emerging from ugly cocoons.

Wells never lost his enthusiasm for a world state ruled by such an elite. In later books he gave this ruling class such names as the "Samurai" (in *A Modern Utopia*, 1904),

"Open Conspirators" (in *The Open Conspiracy*, 1928), and the "Air Police" (in *The Shape of Things to Come*, 1933). Wells's Samurai echo the Guardians in Plato's *Republic*, the first great Utopian and a work Wells greatly admired. Samurai clubs sprang up here and there in England. Wells even made an abortive effort to transform the socialist Fabian Society, then controlled by Sidney and Beatrice Webb, into an organization of "Open Conspirators."

For a short time Wells viewed Lenin's revolution in Russia as not far from his notion of a great state taken over by an efficient elite. When he visited Russia and met Lenin —a visit he recorded in Russia in *The Shadows* (1920)—he found fault with many aspects of communism, but there is no hint that he deplored its total absence of democracy. Indeed, to put it bluntly, the world state outlined here and in *A Modern Utopia* is a police state. Wells never made it clear whether his Samurai would take power gradually or by a bloody revolution. It never occurred to him, one regrets to say, that intelligent, science-trained leaders might be just as willing to work for an evil dictator as for a democratic society that respected free speech and human rights. It was the same mistake that would be made years later by those who called themselves technocrats. Wells seemed never to ask himself why his Samurai, once they controlled a nation, should be more concerned for the welfare of all than the welfare of themselves alone.

In later books and articles Wells would elaborate at length on what he believed was the inevitable emergence of a world state governed by enlightened technocrats. What of the possibility that in the aftermath of wars great nations would fall into the hands of ruthless Napoleons? "Nothing of the sort is going to happen," Wells declares in this book. "The day of individual leaders is past."

As he grew older and wiser, Wells moderated his early attacks on representative democracies in which all citizens have a right to vote. In *The Work, Wealth, and Happiness of Mankind* (1931) he suggests ways of improving democracy by such means as proportional representation, and the elimination of two legislative bodies, in England and America for instance, when a single body would be sufficient. He sees the possibility of civil servants running governments, but admits that "no one has yet worked out any better way [than democracy] of getting general assent to administration and legislation. A much better way cannot be beyond human contriving, but it has not yet been contrived."

Wells also was never clear on exactly how nations would come together to create a United States of the World governed by a science-trained aristocracy that would combine some sort of representative democracy with economic socialism. Was his vision of a worldwide utopia, free forever of wars and gross injustices, a genuine possibility? Or was it no more than a dream?

All his life Wells alternated between moods of hope and moods in which he feared such a world state would never come to pass. As he approached death, a profound pessimism overcame him. He saw no signs of movement toward a world government.

There were only intensifying nationalisms, and hideous wars waged with ever more powerful weapons of mass destruction. Such hopelessness tinged *The Fate of Homo Sapiens* (1939), and culminated in the black despair of *The Mind at the End of Its Tether* (1945). This sad little book was written shortly before Wells died in 1946, but not before he read that atomic bombs, which he had imagined and named thirty years earlier, had fallen on Japan.

Wells's vision of future warfare, the substance of chapter 6 in *Anticipations*, bristles with flawed predictions. Although he sees a role in modern warfare for motor cars and tanks, he calls tanks "iron tortoises," too slow moving to be effective. Modern soldiers, he believes, will move rapidly on bicycles! Rifles, of course, will be improved; guns will have longer ranges. Huge balloons, armed with men, guns, and perhaps bombs, will fight awesome air battles. (Wells devotes several lurid pages to describing such scenes.) And tethered balloons, carrying telescopes, he says, will serve as "argus eyes" to spy on fields ahead, illuminating night landscapes with huge searchlights.

Sea battles will be fought by small, swift ships armed with guns and battering rams. Such battles will be brief. "The struggle between two naval powers on the high seas . . . will not last more than a week or so." Submarines? "I must confess," Wells writes, "that my imagination, in spite even of spurring, refuses to see any sort of submarine doing anything but suffocate its crew and founder at sea." Their torpedoes, he assures his readers, will have as much chance of hitting a target as a man "blindfolded, turned around three times, and told to fire revolver-shots at a charging elephant."

The chapter scores a few hits. The winning side in the modern war, Wells foresees, will be the side with the best educated fighting men, and the best scientists and engineers behind the lines. He correctly perceives that during a great war, an entire nation must turn socialist, its government taking control of every aspect of the economy. He anticipates many useless wars in which young men will be slaughtered for no rational cause.

"Tramp, tramp, tramp, they go, boys who will never be men, rejoicing patriotically in the nation that has thus sent them forth, badly armed, badly clothed, badly led, to be killed in some avoidable quarrel by men unseen." This early in the century, not even Wells could conceive of the deaths of tens of thousands of civilians from bombs dropped from planes. As we know, such senseless wars still rage around the world. Such wars will never cease, Wells was convinced, until there is a viable world state.

Nor would a world government emerge, wells believed, until a worldwide language did. The three major contenders, he writes in chapter 7, are English, French, and German. As other languages, with the exception of Asian ones, fade, Wells sees the western nations slowly becoming bilingual. French will dominate in the short run, but in the long run English will prevail. German is dismissed as too "unattractive, unmelodious, and cursed with a hideous lettering. . . ."

Most of the economic and political arguments of chapter 8 are now hopelessly obsolete. The chapter's central theme is the sluggish movement toward a single government that Wells calls the New Republic. The movement will accelerate as nations become more intertwined economically and culturally, as English becomes the world's dominant language, and as more books are translated into other languages. Wells bemoans the absence of intelligent debates on major issues. He imagines how wonderful it would be if important controversial books were annotated by writers who held contrary opinions. He longs for a general index of all books on sale, listed alphabetically by title and by author. Such an index, though only for American books, is now available in hard cover, and on computer screens as the many-volumed *Books in Print*.

Wells's final chapter swarms with views he would later regret. In it, he reveals that although he considers the omniscient, all-powerful deity of Christianity absurd, he does not want to discard God altogether. In place of the Judeo-Christian God, with infinite attributes, Wells proposes a finite God, unrelated to any established faith. The elite who rule the New Republic will worship such a God, he suggests, recognizing that He is indeed transcendent, as far beyond our understanding as our world is beyond the grasp of an amoeba. Somehow, such a deity will provide a meaning for the universe and human history, even though we cannot know what the meaning is.

Wells seems unaware that this concept of a finite God had earlier been put forth by many philosophers. He would argue again for such a deity in his World War I novel *Mr. Britling Sees It Through*, and in two 1917 works, *God the Invisible King*, and *The Soul of a Bishop*. Later he would repudiate the concept altogether, and declare himself an honest atheist. Today, a finite God is defended by the so-called process theologians, notably Charles Hartshorne, who believe God exists in time, and is evolving as His creation evolves.

Throughout his life Wells vigorously defended a belief in free will. He does not mention Kant but his approach to the free will problem is identical with Kant's. In the world of science, Wells writes in his last chapter, there is strict cause-and-effect determinism. But the human will is somehow outside the "rigidly predestinate" world of "atoms and vibrations." It is free "just as new-sprung grass is green, wood hard, ice cold, and toothache painful." Wells believes that by exercising free will, humanity will finally construct a utopia of peace and justice. The old religious ethics will give way to a new set of morals, though the posits on which such a morality will rest are never made clear.

In the decades before the rise of a world state, Wells predicts that Protestant Christianity will slowly decay, to be replaced, as the world's dominant faith, by Roman Catholicism. Those who refuse to follow this trend, Wells warns, will turn to weird pseudoscientific cults such as theosophy, Spiritualism, and Eastern religions. Rich young men and women will dabble in witchcraft and devil worship just for the fun of it. On this score, of course, Wells's crystal ball was quite accurate. Gilbert Chesterton is thought to have said somewhere that when people stop believing in God, they believe anything. Wells here says it this way: "The fool hath said in his heart, 'there is no God,' and after that he is ready to do anything with his mind and soul."

Wells's passing interest in a finite God was a harmless diversion, but there is a dark side to this chapter: It defends an extreme program of negative eugenics. The New Republic leaders, Wells writes, "favor the procreation of what is fine and efficient and beautiful in humanity—beautiful and strong bodies, clear and powerful minds, and a growing body of knowledge—and to check the procreation of base and servile types . . . of all that is mean and ugly and bestial in the souls, bodies, and habits of men."

Just how is this to be done? By mercy killings! The

leaders of the new world will have "little pity and less benevolence." They "will not be squeamish" about inflicting death on the unfit "because they will have a fuller sense of the possibilities of life than we possess. They will have an ideal that will make killing worth the while; like Abraham, they will have the faith to kill. . . ."

And who are the unfit that are to be thrown away? Wells ticks them off: those with transmittable diseases, with mental disorders, with bodily deformations, the criminally insane, even the incurable alcoholic! All are to be put to death humanely—by first giving them opiates to spare them needless suffering!

How will the New Republic handle what Wells calls "inferior races"? He asks: "How will it deal with the black? How will it deal with the yellow man? How will it tackle that alleged termite in the civilized woodwork, the Jew?"

After the world state is in place, perhaps not until after 2000, Wells foresees that sterilization, killing, and birth control methods will effect a gradual fading from the earth of inferior races. "There is something very ugly about many Jewish faces," Wells writes, then he quickly adds, "there are Gentile faces just as coarse and gross." Many Jews "are intensely vulgar in dress and bearing, materialistic in thought, and cunning and base in method, but no more so than many Gentiles."

Yet Wells sees no reason for any special effort to eliminate Jews. Those who have a tendency toward "social parasitism" will be treated just like similar Caucasians. Increasing intermarriages of Jews and Gentiles, Wells predicts, will be sufficient to cause Jews to "cease to be a physically distinct element in human affairs in a century or so." As for "those swarms of blacks, and brown, and dirty-white, and yellow people," who do not meet the needs of the New Republic, "they will have to go. . . . So far as they fail to develop sane, vigorous, and distinctive personalities for the great world of the future, it is their portion to die out and disappear."

From our perspective, of course, Wells's statements about inferior races, and the use of killing as a tool to weed out the unfit, come perilously close to Hitler's efforts to breed a superior Aryan race, and to "solve the Jewish question" with the aid of gas chambers. Nor do we know whether Wells ever apologized for this portion of his last chapter. In his autobiography (1934) he calls *Anticipations* the "keystone to the main arch of my work." However, as early as 1905, in *A Modern Utopia* (chapter 5) Wells presents a much less harsh defense of eugenics. He still advises ridding the world of the unfit, but now the only means he proposes is sterilization. "So soon as there can be no doubt of the disease or baseness of the individual, so soon as the insanity or other disease is assured, or the crime repeated a third time, or the drunkenness or misdemeanor past its seventh occasion (let us say), so soon must he or she pass out of the common ways of men. . . ." There will be incarceration and sterilization, but "no killing, no lethal chambers."

There is a section on eugenics in a much later work, *The Work, Wealth, and Happiness of Mankind* (1931). Here Wells still purports to favor isolating and sterilizing the unfit, but he now clearly recognizes that human nature is far too complex to justify any controlled breeding of superior bodies and minds. "We do not want human beings to become simply taller or swifter or web-footed or what not. We want a great variety of human beings. . . . We must tolerate much that is odd and weak lest we lose much that is glorious and divine. . . . For many generations, and perhaps for long ages, we must reckon with a population of human beings not very different from those we have to deal with today. . . . It is to a better education and to a better education alone, therefore, that we must look for any hope of ameliorating substantially the confusions and distresses of our present life."

In *Anticipations'* final chapter, Wells suggests that although leaders of the New Republic will believe they are serving an unknowable God, this quasireligious faith will not include a belief in immortality. There may be an after-life, Wells admits, but "on this side, in this life," there is no evidence that our "egotisms" will survive death. Citizens of the world state will replace the notion of heaven with the idea of a bright future for humanity. "For that future these men will live and die." Why they will wish to live and die, especially to sacrifice their life, for a future they will never see, is a question Wells never adequately answers.

Surprisingly, Wells made no changes in the text of *Antic-ipations*' 1914 edition. He did add a new introduction in which he confesses that the book contains "several rash and harsh generalizations," but on the whole he is pleased with "how little there is in it that I would change were I to rewrite it. . . ." His chief apology is for a failure to foresee how rapidly aviation would develop, and for his "very stale . . . anticipations of aerial war." He makes no apology for his "onslaught" on mass democracy, or for his vision of a world state governed by an enlightened, science-minded elite. "It is my faith," he writes. "It is my form of political thought."

EDWARD BELLAMY'S YEAR 2000

> *No man any more has any care for the morrow, either for himself or his children, for the nation guarantees the nurture, education, and comfortable maintenance of every citizen from the cradle to the grave.*

> —Looking Backward, *Chapter 9*

L ooking Backward: 2000–1887, by Edward Bellamy (1850–1898) is far and away the most popular, most influential utopia novel ever written, and also one of the worst. In its endless reprintings it sold over a million copies and was translated into twenty languages. Soon after its appearance in 1888, some hundred books were published either attacking Bellamy's vision of Boston in the year 2000 or defending it. About half of these books

---

This review first appeared in the *New Criterion*, September 2000.

were utopias with such titles as *Looking Ahead*, *Looking Beyond*, *Looking Within*, *Looking Forward*, all even more preposterous than Bellamy's.

William Dean Howells was so taken by *Looking Backward* that he wrote two utopia novels of his own, *A Traveler from Altruria* and *Through the Eye of the Needle*. Both books resemble Bellamy's in their replacement of unbridled capitalism by a moneyless socialist state. England's poet and socialist William Morris was so infuriated by what he called Bellamy's "horrible cockney dream" that he wrote *News From Nowhere* about a less regimented future society with more of an emphasis on crafts than on machinery.*

Looking Backward had an enormous effect on Eugene Debs and later American labor and political leaders who called themselves socialists. In an essay "How I Became a Socialist," Debs thanks Bellamy for "helping me out of darkness into light." The book left its mark on such American socialists and liberals as Norman Thomas, Upton Sinclair, John Dewey, Scott Nearing, Lincoln Steffens, Jack London, Charles Beard, Carl Sandburg, Erich Fromm, and many others. Socialist leaders in England and Europe were also influenced by the book. Leo Tolstoy translated it into Russian.

Vernon Parrington, in the third volume of *Main Currents in American Thought*, devotes fourteen pages to a sympathetic account of *Looking Backward*. David Riesman, in his biography of Thorstein Veblen, tells how Bellamy's utopia altered the lives of Veblen and his wife, and persuaded Veblen to shift his academic training from philosophy to economics.

Engineer Arthur Ernest Morgan, chairman of the Ten-

---

*The most recent fictional spinoffs from Bellamy's utopia are two novels by science-fiction writer Mack Reynolds: *Looking Backward From the Year 2000* (1973), and *Equality in the Year 2000* (1977). Another Julian West goes to sleep and awakes to find a world transformed by nuclear fission energy. Reynolds was an active member of America's Socialist Labor Party, based on the writings of Daniel DeLeon.

nessee Valley Authority under Franklin Roosevelt, and for sixteen years president of Antioch University, thought so highly of *Looking Backward* that he wrote Bellamy's first biography. In *Nowhere and Somewhere: How History Makes Utopias and How Utopias Make History* (1946), Morgan discusses the influence of *Looking Backward* on the shapers of Roosevelt's New Deal, especially on Adolf Berle Jr., whose father was a friend and disciple of Bellamy. "Striking parallels may be drawn," Morgan writes, "between *Looking Backward* and various important aspects of New Deal public policy. It may be said with considerable force that to understand the long range implication of the New Deal one must read *Looking Backward*."

Heywood Broun, during his Soviet fellow-traveling phase and before he converted to Catholicism, wrote an effusive introduction to the Modern Library's edition of *Looking Backward*. Bellamy's utopia, Broun writes, aroused his first interest in socialism. He thinks its description of America in 2000 "is close to an entirely practical and possible scheme of life. . . . There is at least a fair chance that another fifty years will confirm Bellamy's position as one of the authentic prophets of our age."

Oscar Ameringer, known as the Mark Twain of American socialism because of his comic wit, had this to say in his autobiography *If You Don't Weaken*: "Yes, yes, *Looking Backward*. A great book. A very great book. One of the greatest, most prophetic books this country has produced. It didn't make me look backward, it made me look forward, and I haven't got over looking forward since I read *Looking Backward*."

Krishan Kumar, in *Utopia and Anti-Utopia*, quotes John Dewey:

> Bellamy was an American and a New Englander in more than a geographical sense. He was imbued with a religious faith in the democratic ideal. But for that very reason he saw through the sham and pretence that exists or can exist in the present economic system. I could fill

pages with quotations in which he exposes his profound conviction that our democratic government is a veiled plutocracy. He was far from being the originator of this idea. But what distinguishes Bellamy is the clear ardor with which he grasped the *human* meaning of democracy as an idea of equality and liberty, and portrayed the complete contradiction between our present economic system and the realization of human equality and liberty. No one has carried through the idea that equality is obtainable only by complete equality of income more fully than Bellamy. Again, what distinguishes him is that he derives his zeal and his insight from devotion to an American ideal of democracy.

Bellamy was a shy, genial, slender, lifelong New Englander. The son of a Baptist minister, he began his career as a journalist on several newspapers. He became an attorney but never practiced. Several of his early novels and a raft of science-fiction tales are now totally forgotten, but *Looking Backward* became an instant best-seller, then second only to *Uncle Tom's Cabin*. It has never been out of print. At the moment at least seven editions are available in the United States, including a two-dollar Dover paperback.

Looking Backward is narrated by Julian West, a rich, politically conservative young man living in Boston in 1887 and engaged to Edith Bartlett. He suffers from insomnia. Because little noises keep him awake at night he builds a soundproof cellar and hires a mesmerist to put him to sleep. A servant is trained to awake him in the morning. While in a deep trancelike slumber, his house burns down. The basement vault is not discovered and it is assumed that West died in the fire.

Fast forward to the fall of 2000. Dr. Leete, a retired physician, has built a house on West's former property. An excavation revealed the hidden cellar, and within it the perfectly preserved body of West. He has slept for 113 years. This notion of someone entering the future by way of a big sleep had been used many times before—Rip Van Winkle, for example—and would be used again, notably by H. G. Wells in his negative utopia *When the Sleeper Wakes*, and by Woody Allen in his 1973 movie, *Sleeper*.

After being aroused from his hypnotic trance, West becomes the guest of Dr. Leete, his wife, and his daughter Edith. West is attracted at once by the "faultless luxuri-ance" of Edith's figure and her "bewitching" face. The rest of the novel is a detailed account of the brave new world of 2000 in which West is now amazed to find himself.

Dr. Leete has an annoying habit of laughing at almost everything West has to say about nineteenth-century Boston, and Edith blushes at West's slightest remarks. The romance between Julian and this second Edith is embar-rassingly mawkish. It turns out that she is the great-grand-daughter of the earlier Edith, and there is even a hint that she is West's former fianceé reincarnated. Did Bellamy intend "E. Leete" to be a pun, as Everett Bleiler suggests in his massive *Science-Fiction: The Early Years*, on "elite"?

Like Marx's *Das Capital*, which had little influence on Bellamy,* *Looking Backward* gives a fairly accurate pic-ture of the evils of a totally unregulated capitalism. The nation's economy in the late nineteenth century was in the iron grip of gigantic trusts. Graft was rampant in big cities. Labor unions were just getting organized and there were bitter, bloody strikes. The air was dense with coal burning smoke. Anarchists were blowing up buildings.

---

*The books that probably had the greatest impact on Bellamy's socialism were *The Coming Revolution* (1880), and *The Cooperative Commonwealth* (1894) by Danish-born Laurence Gronlund, a promoter of the Socialist Labor Party. The plot of *Looking Backward*, though little else, may have been partly borrowed from a now forgotten utopia, *The Diothas or a Far Look Ahead* (1881) by Ismar Thiusen, pseudonym of John MacNie, professor of French and German at the University of North Dakota. The narrator is projected into the far future by mesmerism where he marries a reincarnation of Edith, his former sweetheart. For a good summary of this novel see Everett Bleiler's *Science Fiction: The Early Years*, pages 734–35.

Everywhere there was enormous wealth alongside slums and miserable poverty.

Julian West's famous allegory of the coach appears in the first chapter of *Looking Backward*. It likens nineteenth-century America to a "prodigious coach" that is being pulled slowly over rough terrain by workers who are harnessed like beasts to a long rope. The coach's driver is hunger. By lashing the workers, hunger forces them to keep pulling.

Riding on the coach in comfortable seats are the rich capitalists. When workers faint from hunger, the riders urge them to be patient. If they become injured or crippled, the rich, out of compassion, give them salves and liniments. They agree it is a great pity that the coach is so hard to pull. The gap between themselves and the poor is rationalized by the "hallucination" that they are made of "finer clay" than the rope pullers.

West's account of the vast changes that have taken place in the twentieth century is, as in all utopias, a bizarre mix of hits and misses. Its most spectacular hit is not its vision of America in 2000, but its description of a command economy that strongly resembles the Soviet Union under Lenin and Stalin, especially their dream of the Communist state they believed would follow a temporary but necessary "dictatorship of the proletariat." It's as if West awoke not in Boston but in Lenin's Moscow!

In the early years of the twentieth century, as Dr. Leete informs West, the great monopolies grew larger and more powerful. No violent worker revolution occurred as in Russia. Unregulated capitalism slowly evolved until the state took over the monopolies and all other means of production to become one monstrous trust, the country's sole capitalist and land owner. The profit motive was gradually replaced by a patriotic desire on the part of everyone to serve the government. Political parties, labor unions, banks, prisons, and retail stores all vanished. In the absence of greed, there was no government corruption.

Even prostitution became extinct. Everybody loved everybody. Crime faded away except for a few unfortunate souls who are mentally ill. They are treated in mental hospitals.

As in the land of Oz, money has totally disappeared. As the nation's sole employer, the government pays no wages. Instead, it provides each citizen with an annual allotment of goods and services. Everyone carries a cardboard credit card that is punched each time a purchase is made or a service rendered.

Edith takes West to a building where samples of all available goods are on display. Their prices, strictly controlled, are in dollars and cents, but these numbers, like algebraic letters, are no more than symbols to aid government accounting. There is no advertising. A buyer tells a clerk what he or she wants, orders are sent through pneumatic tubes to a warehouse, then the goods are shipped through larger tubes to spots from which they are delivered to houses.

There are no household servants. Washing is done in public laundries. Medical and health care is completely socialized. You may choose your doctor. His pay, in the form of goods, is the same as everyone else, including laborers. Public kitchens provide food for home meals, though most people take their main meal at government-run restaurants.

Prizes go to workers for exceptionally good work. If they refuse to do their job properly they are put in solitary confinement on bread and water. Work hours are short and vacations regular. Those too ill to work are placed in "invalid corps" where they do whatever they can. The lame, sick, and blind all receive the same goods as others. There is no longer a division between rich and poor. From each according to his abilities, as a popular Marxist slogan had it, and to each according to his needs.

The state operates like a vast military complex. Every man is conscripted at age twenty-one to serve as a common laborer. At twenty-four all persons, male and female, are given tests to determine their natural aptitudes and wishes. State-run colleges train them for a profession. Some choose "brain work" such as music, art, science, writing, and so on. Workers are free later to change jobs and to live where they like. Retirement is compulsory at age forty-five. October 15 is Muster Day on which those of twenty-four enter the work force, and those of forty-five are mustered out.

All books and newspapers are published by the government, though there is no censorship and one may write anything he or she pleases. Authors pay for first printings. If a book or periodical sells well, the author receives a royalty in the form of goods. Red Ribbons are awarded to outstanding brain workers.

There is no jury system, no attorneys. Legal decisions are made by judges appointed by the president. State governments have vanished. A congress meets once every five years, though just what it does is unclear because new laws are no longer needed. All schools and colleges are run by the state.

What about religion? There are no churches or clergy, no denominations or sects. Persons are free to express their religious opinions in sermons delivered over a telephone system, but the dominant religion is a vague sort of theism, based on the ethical teachings of Jesus to love God and neighbor, much like the deism of the founding fathers.

Fossil fuels have been replaced by electric power that provides heat and light. The air is free of pollution. Chimneys are nowhere to be seen. Wars have become relics of the past, replaced by what William James called its "moral equivalent," working for a better world. All the world's great nations have become command socialisms with universally honored credit cards. There is free trade, free emigration, and the stirrings of a world government. Everyone speaks a native language and a universal language, the nature of which is not specified. Democracy of a sort exists by voting at various government levels. The president, chosen by a small group of peers, serves for five years.

The novel's plot takes a surprising turn at the end. West falls asleep and wakes up back in nineteenth-century Boston convinced that his visit to 2000 was only a dream. He tries to interest others in his dream's utopia only to be ridiculed and thought mad. He then wakes up to find himself back in Boston in the year 2000, the real world, to marry Edith and live happily as a lecturer on the sins of capitalism.

Bellamy never used the word socialism, a term he hated because it suggested European influences and violent revolutions. He called his ideology nationalism. It is hard now to believe, but so persuasive was his rhetoric that over 160 groups called Nationalist Clubs or Bellamy Clubs sprang up throughout the land. There was even a short-lived Nationalist Party. Two journals promoting nationalism flourished for a few years: *The Nationalist* (1889–1894), and Bellamy's own monthly *The New Nation* (1891–1894). There were other periodicals that debated nationalism. The populist movement, embodied in the People's Party, owed a great debt to nationalism.

Not much is said in *Looking Backward* about the role of women in 2000 except that they serve as an auxiliary force in the industrial army under a female General. A much more detailed account of the the status of women appears in *Equality* (1897), Bellamy's lengthy sequel to *Looking Backward*. West continues as narrator. The book makes clear that women enter the work place as equal to men in all respects except for participation in athletic games. Edith, for example, works on a farm. Bellamy supported the suffrage movement, and the early feminist leaders considered him one of their heroes.

In *Looking Backward* Edith dresses in nineteenth-century attire so as not to disturb Julian who is in a perpetual state of shock over everything he sees. In *Equality*, Edith reveals to Julian the clothes she actually prefers. They are men's suits with trousers. Clothes in 2000 are made of paper and discarded after being worn. Also made of paper are shoes, carpets, sheets, draperies, dishes, pots, and pans. When discarded they are recycled to be used in making other things.

Girls, we are told, take over their mother's last name with their father's name as their middle name. Boys do the reverse. Women have free choice over the number of children they desire, though Bellamy is silent about birth control methods and abortions. The pessimism of Malthus has been answered by a stable world population. Blacks are nowhere mentioned in *Looking Backward*. In *Equality*, chapter 37, in a section headed "The Colored Races and the New Order," it is explained that after the freeing of slaves they were soon absorbed into the new order as equal to whites in all respects, although there continues to be no social "commingling" of the two races.

Equality's chapter 19, "Can a Maid Forget Her Ornaments?" (a quotation from Jeremiah 2:32), is pure Veblen. Women no longer wear rings or other costly jewelry to serve as badges of wealth. Since there is no longer a distinction between rich and poor there is no need for such displays. Conspicuous waste has gone the way of the profit motive. When West mentions that in his day persons actually thought that diamonds and other precious stones are intrinsically beautiful, Dr. Leete's reply could have come straight out of Veblen's *Theory of the Leisure Class*:

> Yes, I suppose savage races honestly thought so, but, being honest, they did not distinguish between precious stones and glass beads so long as both were equally shiny. As to the pretension of civilized persons to admire gems or gold for their intrinsic beauty apart from their value, I suspect that was a more or less unconscious sham. Suppose, by any sudden abundance, diamonds of the first water had gone down to the value of bottle glass, how much longer do you think they would have been worn by anybody in your day?

One of Bellamy's most successful predictions is what he calls an electroscope. Although its sounds and images come over telephone cables, it is connected to a world-wide network of wires so that it functions exactly like today's television. Not only does it allow home owners to enjoy music, plays, operas, and lectures, it also permits viewers to see live news events wherever they occur around the world.

Umbrellas have become obsolete because when it rains waterproof canopies are lowered over sidewalks and street corners. Water, electricity, and mail are, of course, free to all households. Horses have been replaced by trains and electric-powered cars. West converses with a young farm lass who is plowing a field with an electric driven machine. Flesh eating has been abandoned for strictly vegetarian diets.

Several chapters in *Equality* describe West's view of Boston and its suburbs from what is called an "air-car." Submarines are mentioned. Of course Bellamy could not conceive of atomic energy, although he does mention that past wars were fought by dropping dynamite from air-cars — the "ghastly dew" in a passage quoted from a Tennyson poem about the future. Nor should we fault Bellamy for not anticipating computers, the moon walk, space probes of planets, and all the other wonder's of twentieth-century physics, chemistry, medicine, and genetics.

There is nothing about psychic phenomena in Bellamy's two utopia novels, but throughout his short life he was fascinated by the paranormal. Extra Sensory Perception is featured in several of his short stories. "To Whom They May Come" tells of an island where natives communicate with each other only by telepathy, a notion that H. G. Wells exploited in his finest utopia, *Men Like Gods*. *Miss Luddington's Sister*, one of Bellamy's early novels, is about Spiritualism.

Always in poor health, Bellamy died of tuberculosis at age forty-eight. Had he lived through the Russian Revolution he would probably have become a dedicated Communist. I do not know whether Lenin or Engels ever actually read Bellamy, although we do know they had only disdain for what they called utopian speculation.

In the final chapter of *Equality* Dr. Leete does his best to counter the main objections that conservatives hurl against socialism: that it opposes religion, stifles incentives, discourages originality, leads to political corruption, violates civil rights, makes everyone behave alike, and so on. To the incentive objection, Dr. Leete's unconvincing reply is that under socialism the old desire to maximize one's wealth is replaced by the higher incentives of doing one's work well and contributing to the common good. As it turned out, in every twentieth-century nation where a government, Marxist or fascist, took total control of the economy, the result was a cruel dictatorship in which not only did the predictions cited above prove accurate, but millions of citizens were needlessly slaughtered.

There are two big lessons to be learned from Bellamy's vision of 2000. First, though admirable in its indictment of unfettered capitalism and in its enthusiasm for building a better world, it projected a cure as bad if not worse than the disease. Bellamy's two books reveal with great starkness how naive and simple-minded were the early socialists both here and abroad. They had no inkling of how socialism would soon come to recognize the power of free-market competition and the baleful results of any effort to eliminate it.

Socialism is, of course, a fuzzy word as impossible to define precisely as capitalism or Christianity. Bellamy's vision has almost no resemblance to the democratic socialism of Norman Thomas, Michael Harrington, Irving Howe, John Kenneth Galbreath, and other leading American socialists, or to any of today's socialist nations in which a vigorous democracy is combined with a mixed economy that is part free market and part government owned or controlled. In the opinion of Milton Friedman and many other conservatives the United States is now a model of democratic socialism. In *Freedom to Choose*, Friedman points out that every plank in

Thomas's platform, the last time he ran for president, has been fulfilled and is now accepted by both Democrats and Republicans. The ability of capitalism to overcome its dark past and become more benign was a development that neither Bellamy nor Marx could foresee.

The second lesson to be learned from Bellamy is that the future is unpredictable, not only with respect to science and technology but also to political and economic change. It's a safe bet that the new millennium will swarm with stupendous surprises that no one now is even capable of imagining. G. K. Chesterton opens *The Napoleon of Notting Hill*, a fantasy set like George Orwell's anti-utopia in 1984, with these wise words.

> The human race, to which so many of my readers belong, has been playing at children's games from the beginning, and will probably do it till the end, which is a nuisance for the few people who grow up. And one of the games to which it is most attached is called "Keep tomorrow dark," and which is also named (by the rustics in Shropshire, I have no doubt) "Cheat the Prophet." The players listen very carefully and respectfully to all that the clever men have to say about what is to happen in the next generation. The players then wait until all the clever men are dead, and bury them nicely. They then go and do something else. That is all. For a race of simple tastes, however, it is; great fun.

POSTSCRIPT

Bellamy's rhetoric reached soaring heights in chapter 23 of *Equality*, titled "The Parable of the Water Tank." The chapter was widely distributed by early socialists, not only here but also abroad and especially in Russia. The parable tells of a dry land where the masses suffer from thirst. Crafty capitalists manage to accumulate large amounts of water which they keep stored in a huge tank (the market).

Workers are hired to bring more water. For each bucket they bring they receive a penny. For each bucket they take from the tank they are charged two pennies.

When the tank overflows, the capitalists stop hiring workers to bring more water. Deprived of wages, the workers no longer can buy water. They start to die of thirst.

The capitalists hire "soothsayers" (economists) to explain the crisis. Some blame it on overproduction, some on sun spots, others on lack of confidence. False "holy men" (preachers) assure the workers that their misery is from God. If they bear their plight in patience they will go, after they die, to a land where there is lots of water.

When some of the workers try to get at the water tank, the capitalists hire warriors (police) to drive them back. Seeing that the people refuse to be pacified, they dip their fingers into the tank and scatter drops of water (charity) on the thirsty.

After the capitalists have wasted all the water by making fountains and fish ponds, and taking frequent baths (Veblen's conspicuous consumption), the tank is emptied, workers are again hired to bring water, and the business cycle begins again. Eventually "agitators" (labor leaders) convince the workers that they should organize and build their own water tanks. When the capitalists realized what was happening, "their knees smote together and they said to one another, 'It is the end of us.' "

And so it came about that there was no more thirst in the land. Everyone called each other brother and sister, and with the greedy capitalists gone, they all live in peace and unity.

AFTERWORD TO
THE SCARECROW
OF OZ

Before writing *The Scarecrow of Oz*, L. Frank Baum had published eight Oz books. Despite their enormous popularity, Baum was always eager to write children's fantasy that had nothing to do with Oz. Seven of his non-Oz novels preceded *The Scarecrow*. The last two, *The Sea Fairies* (1911) and *Sky Island* (1912), originally introduced the protagonists of *The Scarecrow* and two of his best-loved "meat people" from the United States—Trot and Cap'n Bill.

In *The Sea Fairies*, Trot and her peg-legged companion are taken on an undersea tour by three mermaids. We are told that Trot is a nickname for Mayre Griffiths—a wise, brave little girl who lived somewhere on the California coast. She was called Trot because as soon as she could walk she began trotting. Her father, Captain Charles Griffiths, was in charge of a schooner previously captained by

---

This afterword first appeared in L. Frank Baum, *The Scarecrow of Oz* (Kalamazoo, Mich.: 1991).

William Weedles. After Weedles lost his leg, forcing him to retire from seafaring, he lived with the Griffiths, as their "star boarder." It is Cap'n Bill's left leg that is wooden in all of Baum's books, except *Sky Island*. For some curious reason, in that book John R. Neill, Baum's illustrator, puts the peg-leg on the right!

In *Sky Island*, Trot and Cap'n Bill have an even more wonderful adventure in the Pink and Blue Countries above the clouds. They are carried there by the magic umbrella of Button-Bright. This little boy from Philadelphia, who is always getting lost, had earlier been lost and found in Baum's fifth Oz book, *The Road to Oz* (1909).

Trot's family name is spelled Griffith in *Sky Island*. This was either a printer's error or Baum had forgotten his former spelling. In *The Lost Princess of Oz* (1917), we learn that Trot is a year younger than Dorothy. (Another Mayre, from Centerville, is featured in Baum's unproduced musical, *Prince Silverwings* [1903], and anticipates Trot by visiting both sea and sky fairies.)

Two more Oz books followed *Sky Island* before Baum wrote *The Scarecrow*. In his introduction to the latter, he tells us he received many letters from children begging him to admit Trot and Cap'n Bill into Oz. Not only does the book do this, but it also brings Button-Bright back into the company of Trot and Cap'n Bill. Button-Bright, too, becomes a permanent Oz resident.

When Dorothy was allowed to remain in Oz permanently, Baum solved the problem of her separation from much-loved Uncle Henry and Aunt Em by transporting them to Oz with her. We can only guess why neither Trot nor Button-Bright shows any grief over being removed forever from their parents.

We are told in *The Sea Fairies* that, while Trot is away from home, the mermaids stop time for her mother by putting her into a deep sleep, but there is no evidence it was stopped again during Trot's adventures in *Sky Island* or in *The Scarecrow*. Perhaps time goes faster in the skies than on earth because when Trot finally returns home, her mother acts as if Trot had been away for only a few hours. Her absence in *The Scarecrow* is more serious. As Barbara Koelle writes in her perceptive article, "The Tribulations of Trot" (*The Baum Bugle*, winter 1977), neither Trot nor Cap'n Bill "seems to give thought to poor Mrs. Griffiths who must have long since presumed them drowned in the whirlpool. Leaving her in this state seems unkind, and the Royal Historian does not improve matters by having Glinda remark there was 'no way, at present, for them to return to the outside world.' "

"Come, now!" exclaims Koelle. "No Magic Belt? No sorcery of the Wizard or Glinda that can carry them across the desert?"

Trot obviously was unhappy at home. Her father is always at sea, and her mother is a shrew, who doesn't much like Cap'n Bill. Early in *Sky Island*, Koelle reminds us, Mrs. Griffiths gives Trot and Bill such a tongue lashing that it brings tears to the girl's eyes.

Koelle skillfully summarizes the terrible dangers that Trot meets with incredible fortitude in all the Oz books in which she appears. In *The Sea Fairies*, a mermaid gives her a magic gold ring with a pearl setting that will protect her from harm when at sea. Trot still wears the ring in *Sky Island* where it is useless because the mermaids are too distant. In *The Lost Princess of Oz*, when the Wizard notices Trot's ring, she explains that it provides no protection on land. The ring is not mentioned in *The Scarecrow*, but after Trot and Bill are drawn under water by the whirlpool, Trot feels protecting arms around her. She must be wearing the ring because she is obviously being cared for, not by guardian angels but by the "unseen arms" of mermaids.

We learn near the end of *The Scarecrow* that Trot knows a lot about Oz, perhaps having read the Oz books. Also, Button-Bright had told her about Oz in *Sky Island*, though he never mentions having been there until halfway through the story.

Trot's last major appearance in an Oz book is in Ruth Plumly Thompson's *The Giant Horse of Oz* (1928), where, in one episode, Trot is rescued by a merman. Koelle finds the absence of Cap'n Bill in this book "inexplicable."

As for this skipper, Bill Weedles, we know nothing about his relatives, except a brother, encountered in *The Sea Fairies*. Presumably he has no wife or children. A strong bond of love has developed between Trot and the old sailor that comes out in many subtle ways, easily missed on a first reading of *The Scarecrow*. As Koelle points out, the bond is less that of father and daughter than of two equal friends, each with great respect for the other, neither subservient to the other. The pair, separately or together, play important roles in later Oz books by Baum, notably in *The Lost Princess of Oz* and *The Magic of Oz* (1919).

A fascinating article by Dan Mannix, "The Enigma of Button-Bright" (*The Baum Bugle*, Christmas 1972), is my source for most of what follows about the boy. We know he had parents because, at the end of *The Road to Oz*, the Wizard sends him home to them in a huge bubble. Part of his real name (Button-Bright can't remember all of it) is Saladin Paracelsus de Lambertine Evagne von Smith. His home is near Germantown in Philadelphia. We know his parents are wealthy because he is always dressed in expensive clothes. The fashionable sailor suit he has on when Dorothy and the Shaggy Man find him in *The Road to Oz*, Mannix tells us, was known in its day as a "Peter Thompson" after the famous Philadelphia tailor who made and sold it. The house where Button-Bright lived must be large because when he sees Trot's white cottage, in *Sky Island*, he tells her it is "pretty small."

A strange incident involving Button-Bright occurs during Ozma's birthday party in *The Road to Oz*. Santa Claus recognizes Button-Bright as a lad whose home he has visited several times. When Dorothy asks where that home is, Santa laughs and whispers something in the Wizard's ear. The Wizard smiles and nods, but Baum never discloses what Santa said.

The adventures in *Sky Island* were made possible by a magic umbrella that had once belonged to Button-Bright's distant ancestor identified only as an "Arabian Knight." The boy had found it one day while playing in the attic. It was this umbrella that also carried him to Mo where he loses it again ("in the popcorn?" Mannix asks), and we never hear of it again.

"Papa always said I was as bright as a button," the boy tells the Scarecrow at their first meeting, "so mamma always called me Button-Bright." There are all kinds of buttons, the Scarecrow responds, so maybe his father had in mind those buttons covered with dull cloth. "Don't you think so?" the Scarecrow asks mischievously. Button-Bright answers, "Don't know."

Why is the boy considered so dull-witted? Because he is always getting lost, and because he answers so many questions with "Don't know." Baum calls him "stupid" in his introduction to *The Road to Oz*, and Dorothy remarks that Toto "has more sense than Button-Bright," but I agree with Mannix that the boy is far from stupid. His behavior is always that of an intelligent, polite, self-possessed little boy who calmly accepts whatever happens. His "don't knows" suggest that he doesn't want to bother answering, or that he is too honest to say he knows something when he doesn't. In *Sky Island*, when Trot says that Cap'n Bill knows everything, Button-Bright disagrees. "I know some things Cap'n Bill don't know. . . . Some folks think I'm stupid. I guess I am. But I know a few things that are wonderful."

It is the foxy King Dox of Foxville, in *The Road to Oz*, who recognizes that behind Button-Bright's humble "don't knows" is a "brilliant mind." When King Dox asks if the boy knows why two and two make four, he is rightly impressed by Button-Bright's "No." "Clever! clever indeed," raves the king. "Of course you don't know. Nobody knows why; we only know it's so, and can't tell why it's so. Button-Bright, those curls and blue eyes do not go well with so much wisdom. They make you look too youthful, and hide your real cleverness." With a wave of his paw, the king turns Button-Bright's head into the head of a fox so he will look as intelligent as he really is. If you read carefully everything Button-Bright says, observe his curiosity about how things work and his efforts to learn, and consider the many bright suggestions he often makes, you will understand how perceptive King Dox was.

Like Trot, Button-Bright also has no strong desire to go home. In *Sky Island*, he speaks of being raised by a "cross" governess and in *The Road to Oz*, we learn that no one had ever told him much about "fairies of any kind." After losing his magic umbrella in Mo, he tells Trot he doesn't belong anywhere. Without the umbrella, he adds, "it stands to reason that if I can't get back, I haven't any home. But I don't care much. This is a pretty good country, Trot. I've had lots of fun here."

At Ozma's birthday party in "The Road to Oz," the Scarecrow asks, "Don't you want to find your mamma again?" Button-Bright's forthright reply is "Don't know." Although he tells Santa he wants to return home in one of the Wizard's magic bubbles, after Dorothy asks if he is glad to leave Oz, he again answers, "Don't know." We are forced to assume that the boy was so neglected by his wealthy parents that, when Ozma tells him in *The Scarecrow* that he can remain forever in Oz with Trot and Cap'n Bill, he is happy to accept.

In the first Oz book, *The Wonderful Wizard of Oz* (1900), a cyclone blows Dorothy to Oz. In *Ozma of Oz* (1907), a storm at sea takes her to Ev, a magic country separated from Oz by the Deadly Desert. Another shipwreck sends Oklahoma's Betsy Bobbin to Oz in *Tik-Tok of Oz* (1914). The great San Francisco earthquake drops Dorothy into an underground fairyland in *Dorothy and the Wizard in Oz* (1908). After the whirlpool starts Trot and Cap'n Bill on their adventures in *The Scarecrow*, Baum apparently ran out of disasters as devices for getting American children to Oz. At any rate, his later Oz books all take place entirely in Oz or in nearby magic lands.

Phunnyland, where it rains lemonade and snows hot buttered popcorn, was the setting for Baum's first non-Oz fantasy novel, *A New Wonderland* (1900). Three years later, the name of the country was changed to Mo, and the book was reissued as *The Magical Monarch of Mo*. I suspect that Jinxland was put inside the Quadling region of Oz to make *The Scarecrow* seem more like an Oz book rather than the third story Baum had been planning for a series about Trot and Cap'n Bill. Without Oz in a book's title, Baum's fantasies did not sell well, but if he ended a non-Oz fantasy by bringing his characters to Oz at the last moment—as he did in *Dorothy and the Wizard in Oz* and *Rinkitink in Oz* (1916)—he could justify Oz in the book's title.

The Scarecrow of Oz is a misleading title because the Scarecrow is not a main character. Not only does he play a minor role in the novel's last half, but even his great plan to defeat King Krewl "fizzles," as the Ork shrewdly predicted. The straw man's brains are seldom as good as he thinks they are. He seems to have had no plan at all other than to beat the king with a whip, which results in the Scarecrow's immediate capture and almost permanent destruction by fire.

We must not forget the Ork, for he not only saves the Scarecrow from certain doom but is one of Baum's most bizarre animals. His name is Flipper, and he lives in Orkland, an island in the Nonestic Ocean, east of Oz and not too far from the Enchanted Island of Yew and the inland Valley of Mo. The Ork's propeller, which can turn both ways (enabling him to fly backward?), is not the first example in fantasy of an animal with an organic part that rotates. The Wheelers, in *Ozma of Oz*, have wheels instead of hands and feet. Amazingly, it was discovered in the mid-1970s that bacteria swim by actually rotating their flagella like propellers! (See "How Bacteria Swim," by Howard Berg, in *Scientific American*, August 1975.)

There has been much speculation about Baum's decision to call the large bird an Ork. Ruth Berman, writing in

Mythlore (January 1969), discloses that medieval writers used the orc (sometimes spelled ork) for any sort of sea monster. In *Paradise Lost*, Milton writes (book 11, lines 833–35):

> Down the great River to the opening Gulf,
> And there take root, an island, salt and bare,
> The haunt of seals, and orcs, and sea-mews clang—

Baum may have known these lines, German conjectures, and mistakenly thought the orcs, in Milton's line, were birds like the sea-mews (gulls), who clang (make noise). Berman also finds orcs in *Beowolf*; in Tolkien's *Lord of the Rings*, where they are cruel humanoid monsters; and in Ariosto's *Orlando Furioso*, where orcs are sea monsters off the coast of Ireland. In Germanic legends, the Orkneys are whirlpools. It has also been observed that o-r-k are the last three letters of "stork," and that transposing the first two letters of "ork" produces "rok," suggesting the mythological giant rocs in the *Arabian Night* saga of Sinbad the Sailor.

The style of writing in *The Scarecrow* is vintage Baum, simple and unadorned. Critics like to fault Baum for not being a great stylist, like E. B. White (*Stuart Little*, *Charlotte's Web*, etc.), but Baum well knew that children are bored to death by fancy writing. Colorful metaphors and poetic descriptions of scenery and weather are only impediments for children eager to get on with the action.

Baum's love of puns and other forms of word play are held in check, for the most part, in *The Scarecrow* except for their use in proper names. Pessim, who always sees the worst side of everything, is an obvious play on "pessimist." King Krewl is cruel, King Phearse is fierce, and King Kynd is kind. Button-Bright uses a turkey leg as a drumstick to beat a drum. Trot's erudite pun on "bunion" is surely not understood by today's children (although perhaps by children of 1915), maybe not even by today's adults.

"Corns? Nonsense! Orks never have corns," protested the creature, rubbing its sore feet tenderly.

"Then mebbe they're—they're—What do you call 'em, Cap'n Bill? Something 'bout the Pilgrim's Progress, you know."

Did Baum have anything in mind when he chose the name Pon? Ruth Berman has suggested (in a letter) that he may have been searching for a nickname similar to Jon, Ron, and Don, but slightly exotic. "Pond," she adds, "is an appropriate name for part of a garden. And Pon, it occurs to me, is hidden in the opening line of so many old fairy tales, 'Once upon a time. . . .' "

Lord Googly-Goo's name derives from two slang expressions of the late 1880s. "Googly eyes" referred to big, round, protruding eyes; to "make goo-goo eyes" meant to look at someone in a flirtatious, seductive way. In 1923, a hit song, with words and music by Billy Rose, was written about the comic strip character Barney Google "with his goo-goo-googly eyes." No doubt the rich, detestable Googly-Goo made as many goo-goo eyes at Gloria as Gloria made at Pon. Fred Meyer called my attention to the initials F. R. G. concealed in Googly-Goo's ruff in Neill's illustration on page 143. They stand for Neill's friend Frederick R. Gruger. Neill similarly concealed initials and names of friends in other books by Baum, although you may need a magnifying glass to find them.

I had to look up "trump" (on page 261) in the dictionary to find that it means "good fellow," or (as *Webster's Collegiate Dictionary* has it, not ignoring the ladies) "a dependable and exemplary person." The slang term no doubt derives from the value of a trump card. Though not slang, "fustian" also gave me pause. It means the rough cotton or linen fabric similar to sackcloth, forced on King Krewl as penance in chapter 20. "Give molasses" (on page 218) was a common expression in Baum's day for the tendency of grasshoppers to squirt a sticky substance from their mouths. "Spit tobacco," Mannix writes, was the phrase he knew for this when he was a boy. Occasional slang expressions seem curiously up to date such as the Ork's "Wow!" in chapter 2.

There are touches of Carrollian nonsense here and there. Pessim recommends holding an umbrella over fish during a rain so they won't get wet. The Ork insists that his feet hurt four times as much as Cap'n Bill's, rather than just twice as much, because Bill has only one real leg. The lavender berries that make one shrink, and the purple berries that enlarge, remind readers of the Blue Caterpillar's mushroom in the first *Alice* book. Nibbling one side made Alice shorter; nibbling the other made her taller. Baum has as much fun as Lewis Carroll in altering the sizes of Trot, Bill, the Ork, and the birds of Mo.

The Scarecrow is one of only two Oz books—the other is *Tik-Tok of Oz*—that features an operetta-like boy/girl romance. This is not surprising, as they are the only two Oz books based on earlier dramatizations. The *Tik-Tok Man of Oz*, a 1913 musical comedy, served as the basis for the Oz book of the following year. And the romance and many other aspects of *The Scarecrow* were foreshadowed in a motion picture written and directed by Baum for his Oz Film Manufacturing Company. Titled *His Majesty, the Scarecrow of Oz*, it was produced in 1914, a year before *The Scarecrow of Oz* was published. Dorothy (played by Violet MacMillan) was the heroine. Trot is not in the film, but King Krewl, Googly-Goo, Pon, and Gloria are there. Button-Bright was also in the film, played by Mildred Harris—an Oz Film Manufacturing stock player and soon to be Mrs. Charlie Chaplin. Old Mombi, lifted from *The Marvelous Land of Oz* (1904), is the wicked witch, and the Scarecrow, Tin Woodman, Cowardly Lion, Wizard, and Sawhorse also play roles.

For a summary of the film's wild plot, see "The Oz Film Manufacturing Company, Part 2," by Richard Mills and David L. Greene (*The Baum Bugle*, spring 1973). It is Mombi who freezes Gloria's heart and also turns Pon into a kangaroo. The Wizard finally stuffs Mombi into a large can labeled "Preserved Sandwitches," shrinks it to a small size, then paints out the letters s-a-n-d and e-s in "Sand-witches," imprisoning Mombi. She is freed only after agreeing to unfreeze Gloria's heart and to restore Pon. All of these scenes are shown using 1914 special effects. More effects create the horror scene in which the Tin Woodman chops off Mombi's head. After hunting about, the witch finds it, reattaches it, and goes on her way.

Fortunately for its readers, *The Scarecrow of Oz* is a much better crafted work than the film that preceded it. Indeed it was one of Baum's own favorites as well as that of many of his readers.

In a heart-rending mini-essay in *The Baum Bugle* (Christmas 1960), writer William Lindsay Gresham—best known for his novel about carnival life, *Nightmare Alley*—recalls an occasion when *The Scarecrow* got him through a night of intense despair. His wife Joy had divorced him and left the country with their two children to settle in England and eventually to marry the famous writer and Christian apologist, C. S. Lewis. Gresham's house had been sold, and he was driving alone to New York City. Unable to face a hotel room, he telephoned friends and asked to stay with them.

Struggling to sleep in the room of a daughter who was away at school, not sure he could survive until morning, Gresham spotted a copy of *The Scarecrow* on a shelf. It was the first Oz book he had owned. A grandmother had read it to him many times and, although he later acquired all of Baum's Oz books, this was always his favorite.

> Let no one doubt the magic of Oz, for with the first sen-
> tence it stole from the pages like a warm, enfolding shawl
> about the shoulders of a half-frozen wayfarer. Beneath the
> acacia tree on the cliff I stepped back through the time-
> warp to join Trot and Cap'n Bill on the start of the adven-
> ture which landed them in Jinxland, to take part in the
> lives of the Princess Gloria and Pon, the gardener's boy.

As I read I realized for the first time how powerfully the Oz chronicles had influenced my life, how many healthy and sturdy values I had gained from Baum. I thrilled to the Scarecrows heroic facing of the bonfire and my perspective was restored by his disposal of the wicked witch, Blinkie, and King Krewl—victory without vengeance.

Then came a great healing insight, healing for the wound every child suffers when he is told that Oz is not True. It came like a triumphal chorus, an Ode to joy, that Oz *is* True—every golden word of it It is not fiction but myth—which states greater truths than mathematical formulae can ever encompass, for it glows with the immortal truth that life moves onward and upward, around obstacles and through terrors toward the sun. Oz speaks to us in the primordial images—wicked witches, loving water sprites, the cavern of the sea, the helpful animals (certainly the Ork is champion of these); they are the archetypes of thought, the language of the soul of man.

At several places in the story I gave way to tears of gratitude that a world of sanity was returning to me through the remembered pages. I fell asleep before I had quite finished the book but it didn't matter. I knew it by heart anyhow.

What mattered was that once again, and this time in life, as well as in art, that ever-dependable friend, the Scarecrow, had crossed a bottomless gulf—despair—on the magic cobweb of Ozma's love and Baum's genius to thaw a frozen heart. Not the Princess Gloria's this time, but mine.

INTRODUCTION TO
THE MARVELOUS LAND
OF OZ

After the spectacular success in 1900 of his *Wonderful Wizard of Oz*, one would have expected Lyman Frank Baum to have started at once on a sequel, but if he had plans to write a second Oz book he was in no hurry to begin. His next fantasy, *Dot and Tot of Merryland*, 1901, was lavishly illustrated by William Wallace Denslow, who had illustrated *The Wizard*, and that same year also saw the publication of Baum's *American Fairy Tales*, a collection of unrelated short stories, and *The Master Key*, his only attempt at a science-fiction novel. In 1902 he published *The Life and Adventures of Santa Claus*; in 1903, *The Enchanted Island of Yew*.

As Baum explains in his preface to *The Marvelous Land of Oz*, it was the great stage success of *The Wizard* in 1902 and the thousand letters from children that finally per-

---

This introduction first appeared in L. Frank Baum, *The Marvelous Land of Oz* (New York: Dover, 1969).

suaded him to write a second Oz book. His original title was *The Further Adventures of the Scarecrow and the Tin Woodman.* Played respectively by Fred Stone and David Montgomery, the Scarecrow and the Tin Woodman were the stars of *The Wizard* musical which, as late as 1904 when the second Oz book appeared, was still on the road and playing to capacity audiences. The book is dedicated to the two comedians and the end papers of early editions bore photographs of them as they appeared in the operctta. The book was advertised in *Publisher's Weekly* as *The Further Adventures of the Scarecrow and the Tin Woodman,* but just before it went to press it was decided that Oz should be in the title. The first edition was issued in 1904 by Reilly & Britton, a Chicago house that later became Reilly & Lee. In later editions the title was shortened to *The Land of Oz.* This Dover edition, like its companion *The Wonderful Wizard of Oz,* is an unabridged reprint of the first edition, complete with the original line drawings and color plates.

By 1904 Baum and Denslow had come to a parting of ways, and Reilly & Britton had chosen John Rea Neill, a twenty-seven-year-old artist then living in Philadelphia, to illustrate the new Oz book. Neill remained the Royal Illustrator of Oz for all of Baum's subsequent Oz books as well as for many of his non-Oz fantasies and for all nineteen of the Oz books by Baum's worthy successor, Ruth Plumly Thompson. Neill even wrote and illustrated three Oz books of his own, and wrote a fourth that was never published. Children loved his Ozzy interpretations of Baum's characters and enjoyed searching his pictures for touches of Irish whimsy. In the color plate opposite page 109 in this edition there is a check, barely readable on the floor of the Royal Treasury, made out to Neill for no less than $100,000,000,000.

Baum apparently had little interest in writing more about Dorothy Gale. She plays no role in the second Oz book, although three years later, yielding to the demands of readers, she became the heroine of his third Oz book and appeared in every later Oz book that he wrote. *The*

Marvelous Land of Oz is the only Oz book by Baum in which there is no character from the outside world, although at one point in the story the adventurers find themselves beyond the borders of Oz, just across the southeast corner of the Deadly Desert. (According to the latest official map of Oz and its environs, by James Haff, cartographer, and Dick Martin, artist, they were on a mountain top east of Mo, between the countries of Aurissau and Ribdil.) The Cowardly Lion and the Wizard are also absent from the story. But Glinda the good witch of the South, is back; and we meet again the Soldier with the Green Whiskers, the Guardian of the Gates, the Queen of the Field Mice, and Jellia jamb ("Jelly or Jam"), the pert little palace housekeeper who later became Ozma's Maid of Honor. Jellia is mentioned in the first Oz book but not given a name and is the heroine of Ruth P. Thompson's *Ozoplaning with the Wizard of Oz* (1939).

The Ozziest of the new characters are Jack (for "Jack-o'lantern") Pumpkinhead, the Saw Horse, and the Woggle-Bug. Good-natured Jack, with his eternal grin, is not nearly as seedy-headed as he sometimes seems. The Royal Historian errs at the close of the book when he speaks of Jack as living happily "to the end of his days." No one dies naturally in Oz. We learn in *The Road to Oz* that Jack becomes a prosperous Winkie farmer who raises pumpkins. As soon as his head begins to spoil he simply carves himself a new one and plants the old. He is the central figure in Ruth Thompson's *Jack Pumpkinhead of Oz* (1929).

The Saw Horse—the most famous wooden horse in fantasy literature since Clavileño, in chapter 40 of the second book of *Don Quixote*—remained in the Emerald City, his legs shod with gold, to draw Ozma's Red Wagon. In *The Emerald City of Oz* the Wizard provides him with some excellent sawdust brains made from knots of wood to enable the creature to think out knotty problems.

Professor H. M. Woggle-Bug, T. E., is Baum's caricature of the pompous, overeducated pedagogue, addicted to big words, Latin phrases, and dull puns. One of his puns in this book (on "horse-and-buggy") is so horrendous that even the kindly tin man threatens him with his axe. There is a pleasant discussion of puns on pages 160 and 206. Opportunities to make them are so great, the Woggle-Bug argues, that anyone of genuine culture and refinement is unable to avoid them. In a later Oz book Professor Woggle-Bug becomes president of the College of Art and Athletic Perfection, having invented a pill that provides instant knowledge when swallowed. This frees students from the drudgery of attending classes and gives them more time to concentrate on sports.

In 1905 Baum tried to repeat the stage success of *The Wizard* by writing another operetta, *The Woggle-Bug*, based on his second Oz book. After a trial run in Milwaukee it opened at the Garrick Theatre in Chicago on June 18, 1905, but ceased to woggle, as one reviewer put it, a few weeks later. Although Baum followed the book's plot line closely—too closely, most reviewers thought—he introduced a number of novel stage spectaculars. The Army of Revolt, for instance, bombarded the Emerald City with balloon cannon balls. There was something called the "rain of cats and dogs" that critics did not care for, but they liked the clever way that animated drawings, projected on a screen, showed how the Woggle-Bug escaped from the screen in his highly magnified state. Fred Mace, later a star of Mack Sennett comedies, played the Woggle-Bug, singing such songs as "There's a Lady-Bug A-Waitin' for Me," written by Baum and set to music by Frederic Chapin.

The operetta was publicized by the publication of a large picture book in paper covers, *The Woggle-Bug Book*, in which Baum tells of the Bug's adventures in an unnamed American city. It is a mediocre tale, although the book itself is now one of the scarcest of Baum items. The musical and picture book were also promoted by a series of newspaper comic strips called *Queer Visitors from the Marvelous Land of Oz*. The series started in the Sunday "funny paper" pages of the *Chicago Record-Herald* and other newspapers on August 28, 1904, and continued for six months. The episodes, illustrated by Walt McDougall, deal with adventures of the Woggle-Bug and his companions as the Gump carries them to various American cities. For the first four months each story ended with the Woggle-Bug making an unprinted remark in the last panel. Money prizes were offered for the best answers to the question, "What did the Woggle-Bug say?" The contest was promoted by large newspaper advertisements and posters throughout Chicago, and by thousands of buttons asking "What did the Woggle-Bug say?" Some of these stories, rewritten for small children and colorfully illustrated by Dick Martin, were published by Reilly & Lee in 1960 as *The Visitors from Oz*. (For these details on the operetta and its promotion I am indebted to Frank Joslyn Baum's and Russell MacFall's biography of L. Frank Baum, *To Please a Child*, Reilly & Lee, 1961.)

There is little remarkable about the Gump except its lack of a body, its glass eyes (acquired of course when the head was stuffed), and its curious name. According to *The Dictionary of American Slang*, edited by Harold Wentworth and Stuart Berg Flexner, from 1865 to 1920 "gump" was a slang word for a stupid fellow. This may explain Baum's choice of the word as well as its later use in the famous comic strip about Andy Gump. How the Gump originally came to be killed is something of a mystery, for life is difficult to destroy in Oz. Such destruction rarely occurs, and can be accomplished only in extremely unusual ways.

Tip, the most important flesh-and-bone person in the story, appears in Neill's pictures as a robust lad, although we are told (page 39) that he was "small and rather delicate." This is Baum's only hint of the big surprise to come. It is amusing to note that in the first issue of a small printed newspaper called the *Ozmapolitan* (it is reproduced in *The Baum Bugle*, spring, 1963), put out by Reilly & Britton in 1904 to publicize the second Oz book, a column headed "city news in brief" has two items about Tip even though Ozma has already begun her reign and is mentioned in the same column! Jack Snow, who wrote two Oz books and *Who's Who in Oz*, was so haunted by the separate personalities of Tip and Ozma that he wrote a curious short story, *A Murder in Oz* (it first appeared in *The Baum Bugle* and is reprinted in *The Best of the Baum Bugle*, 1966), in which Tip is actually restored to existence as a separate identity.

Jinjur, the young dairy maid who is so "full of ginger," is Baum's portrayal of a suffragette of the time. The description of her on page 85 is that of a typical feminist, albeit prettier than most; and although her attempt to overturn a strawman and become the female dictator of Oz is eventually thwarted, she retains her masculine protest traits throughout the later Oz books. We learn in *Ozma of Oz* that she has married and has a habit of beating up her husband, and in *The Patchwork Girl of Oz* we are told that one of her jobs is to repaint the face of the Scarecrow when it becomes faded. She becomes such a skillful painter that on one occasion, recalled by the Scarecrow in *The Tin Woodman of Oz*, when the strawman was in need of fresh straw, Jinjur painted such a realistic picture of a haystack that he simply helped himself to the hay, finding it of excellent quality. Baum's satire on the feminist movement is most explicit on pages 170–71, where he describes the men taking over their wives' chores, and on page 283 where he reports the joys of both sexes after the status quo is finally restored.

Mombi, stripped of her evil powers, becomes the most troublesome ex-witch in Oz until she is finally melted by the Scarecrow and Sir Hocus of Pokes in Ruth Thompson's *Lost King of Oz*; (1925). (The "lost king," by the way, is Pastoria, Ozma's father, mentioned here on pages 240–41 and believed to be dead.) Dr. Nikidik, the crooked magician, was reported (in *The Road to Oz*) to have been killed by falling down a precipice, but in *The Patchwork Girl of Oz* he reappears as the crooked Dr. Pipt, living in the Blue

Forest of the Munchkin country and still making his marvelous Powder of Life. Eventually Dr. Pipt also is deprived of his magic powers and after his crooked limbs are straightened by the Wizard becomes a rather decent man. As Lee A. Speth pointed out in *The Baum Bugle* (spring, 1965), Dr. Nikidik obviously had spread a false rumor about his death, changed his name, and moved to the Blue Forest where he and his wife Margolotte escaped detection until the incidents of *The Patchwork Girl of Oz* took place.

Professor Nowitall ("know it all"), from whom the Woggle-Bug acquired his vast erudition, does not reappear in any later Oz book. This is hard to understand because we are told that he is the most famous scholar in Oz; no doubt he is the author of the *History of the Land of Oz* that we see in the illustration on page 151, a history mentioned in several subsequent Oz books.

Like all great writers of fantasy, Baum achieved his strong suspension of disbelief by careful attention to little details. Jack Pumpkinhead, for example, is unable to whistle in one scene because he cannot purse his carved lips, and in another scene the Scarecrow cannot swallow one of Dr. Nikidik's wishing pills because he cannot open his painted mouth. He cannot even pick up a pill with his clumsy padded fingers. Baum was skillful in thinking of such touches but often careless with respect to details that later slipped his mind. Ozma has gold hair in this book; when she next appears, in *Ozma of Oz*, she unaccountably has become a brunette. We know that grass acquires the coloration of the country because it changes from purple to green when Tip and Jack pass from the Gillikin region to the Emerald City, yet on page 237 we are told that Glinda's palace has a velvety green lawn. Perhaps Glinda imported grass from the Emerald City or used her sorcery to alter its color.

Glinda's threat to kill Mombi does not, however, contradict the fact that no one can be killed in Oz, for we know that witches are exceptions. After all, in the first Oz book one witch was destroyed when Dorothy's house fell on her, and another by the water Dorothy tossed from a pail. When Mombi, in the form of a Griffin, ran onto the Deadly Desert she should have turned instantly into dust, but perhaps this is another respect in which witches are exceptions to the natural laws of Oz. We must remember, too, that the Royal Historian acquired more accurate information about Oz as his series progressed, often correcting earlier misinformation or supplying facts about which he was previously ignorant. In the first book, for instance, he did not know the name of the Gillikin region or that its dominant color was purple.

In *The Marvelous Land of Oz*, as in all of Baum's fantasies, there are levels of humor and philosophy that younger readers fail to understand or only sense dimly. The fun poked at the feminist movement meant little to Baum's child readers, though of course it could not have been missed by their elders. The book's funniest scene is surely the first meeting of the Scarecrow and Jack Pumpkinhead. Each has the preconceived notion that they speak different languages. They discuss their inability to communicate, finally calling in Jellia Jamb to act as interpreter. The mischievous little maid deliberately mistranslates, producing as much confusion as a United Nations debate between the United States and the Soviet Union, before it finally dawns on Jack and the straw man that they are talking the same language. Children can appreciate the absurdity of this situation, but I suspect that Baum was aware of metaphorical depths. All men share a common human nature, in spite of our prejudices that make other races and cultures appear barbaric, and in spite of persistent efforts by cultural relativists to deny that there are universal human values.

The book contains Baum's usual touches of Carrollian nonsense. The Saw Horse, just brought to life, doesn't know that "whoa" means "stop" until Tip explains that the two words are synonyms. Jack is unable to see purple because everything is purple. The penalty in Oz for chopping leaves from the royal palm tree is to be killed seven times, then imprisoned for life. There is the difficulty in following Dr. Nikidik's wishing-pill directions to count to seventeen by twos, and the strange time reversal that occurs when Tip wishes he had never swallowed the first pill. Its return to the box is followed by the disappearance of a stomach pain which Tip never actually had since the pill was never swallowed, although Tip insists it was a "splendid imitation of a pain."

Notes of existential wonder sound in the Saw Horse's astonishment at finding himself alive; and when the Scarecrow says (page 155) that "everything in life is unusual until you get accustomed to it," he is expressing a way of looking at things that is a dominant theme in the writings of G. K. Chesterton. The conflict between head and heart, so pervasive in the first Oz book, returns in this sequel in several spots. We are reminded (page 286) that a good heart cannot be created by brains or bought by money. It is the man of brains, the Scarecrow, who asserts (page 188): "I am convinced that the only people worthy of consideration in this world are the unusual ones. For the common folks are like the leaves of a tree, and live and die unnoticed."

"Spoken like a philosopher!" exclaims the Woggle-Bug.

This despicable view, indeed defended by many philosophers, had earlier been countered by the Tin Woodman, of the hollow head but great heart, in a remarkable, perhaps unintended, double pun (heart, tart, tart). "A good tart," the tin man informs Jack Pumpkinhead (page 182), "is far more admirable than a decayed intellect." Upon reflection, the remark expresses an essential aspect of both religion and democracy.

Two motion pictures have been based on *The Marvelous Land of Oz*. The first was written and produced in 1908 by Baum himself as one of a number of hand-colored films that were projected onto a screen while Baum stood at one side to provide dialogue. Selig Pictures released it in

1910 as a one-reeler. General Jinjur and her revolution provided the main plot, but the story also included Dorothy, the Cowardly Lion, and the Wizard.

Half a century passed before the second film version, "The Land of Oz," an hour-long adaptation by Frank Gabrielson. It opened the Shirley Temple Show on NBC color television Sunday evening, September 18, 1960, and was shown again the following year. Tip was played by Shirley Temple herself. Ben Blue was the Scarecrow and Gil Lamb the Tin Woodman. Jack Pumpkinhead was played by Sterling Holloway, Mombi by Agnes Moorehead, Glinda by Frances Bergen. The Saw Horse was an animated wooden figure whose mouth opened and shut when he talked. There was no Woggle-Bug and no General Jinjur with her army of "revolting" girls. Jonathan Winters played the heavy as Lord General Nikidik, the wickedest man in Oz, with Arthur Treacher as his trusted servant. Nikidik and Mombi together plan the capture and transformation of Ozma into Tip. After Nikidik and his army are finally conquered by Glinda, his punishment is to be forced to do one good deed every day and hereafter be known as Nikidik the Good. There were many dull departures from the book but on the whole it was an amusing hour. Particularly good was a scene in which Tip and Jack ask directions of a man, with a jeweler's glass in his eye, who is busy repairing the light of a firefly. The man's official job is to keep all Oz fireflies in good working order.

INTRODUCTION TO *THE BALL AND THE CROSS*

*"Think not that I am come to send peace on earth:
I came not to send peace, but a sword."*
—Matthew 10:34.

T he Ball and the Cross was G. K. Chesterton's second
novel and like all his novels is a mixture of fantasy,
farce, and theology. The basic plot is easily summarized.
Evan MacIan is a tall, dark-haired, blue-eyed Scottish High-
lander and a devout but naive Roman Catholic. He is so
politically conservative that he is a Jacobite who longs for
a restoration of the Stuart monarchy. James Turnbull is a
short, red-haired, gray-eyed Scottish Lowlander and a
devout but naive atheist. Politically he is a romantic
socialist. (In his previous novel, *The Napoleon of Notting
Hill*, G. K. gave the name Turnbull to the owner of a

This introduction first appeared in G. K. Chesterton, *The Ball and the
Cross* (New York: Dover, 1995).

curiosity shop who is also a military genius.) Both men are fanatical in their opinions.

The two meet when MacIan smashes the window of the street office where Turnbull publishes an atheist journal. This act of rage occurs when MacIan sees posted on the shop's window a sheet that blasphemes the Virgin Mary, presumably implying she was an adulteress who gave birth to an illegitimate Jesus. When MacIan challenges Turnbull to a duel to the death. Turnbull is overjoyed. For twenty years no one had paid the slightest attention to his Bible bashing. Now at last someone is taking him seriously! Most of the rest of the story is a series of comic events in which the two enemies wander about seeking a spot for their duel, only to be forever prevented from fighting by the police and other civil authorities.

Two asymmetrical romances enter the narrative. MacIan the believer falls in love with Beatrice, an English-woman who is an agnostic. Turnbull, the skeptic, is smitten with Madeleine, a Frenchwoman who is a prac-ticing Catholic. At the end, each man comes to admire the other for the courage of his convictions, and they become good friends even as they continue their efforts to fight. At one point MacIan saves Turnbull from drowning.

On the lowest level of allegory Chesterton is celebrating in this story a vigorous verbal duel that he had fought with Robert Peel Glanville Blatchford, a now-forgotten but then widely read journalist who was an archenemy of Chris-tianity. In book after book, article after article, Blatchford exalted atheism, blasted the Bible, and trumpeted socia-lism. Chesterton considered him a worthy opponent, and for two years the two men, as G. K. put it in a letter, crossed "intellectual swords."

Blatchford attacked Chesterton in his own socialist weekly, *The Clarion*, and G. K. slashed back in his column in *The Daily News*, in the Christian socialist periodical *The Commonwealth*, and in three articles for *The Clarion* that you will find reprinted in *Collected Works of G. K.*

Chesterton, vol. 1 (Ignatius Press, 1986). Dozens of the arguments Chesterton used with gusto in *The Clarion* are later repeated by MacIan and in Chesterton's *Orthodoxy*.

Chesterton of course bore no outward resemblance to MacIan, who was tall and thin, and totally lacking G. K.'s sense of humor. Nor did Blatchford have any physical resemblance to Turnbull. In his autobiography Chesterton describes Blatchford as "an old soldier with brown Italian eyes and a wondrous mustache." There is no doubt, however, that *The Ball and the Cross*, which G. K wrote only a year after his battle with Blatchford, is a comic portrayal of that battle with real swords taking the place of verbal thrusts.

Just as almost no one in Chesterton's novel takes seriously the duel between MacIan and Turnbull, so almost no one in England paid serious attention to the Blatchford-Chesterton debate. Amusing, yes; significant, no. Even Blatchford said he thought G. K. was a mere actor pretending to be a Christian. In his memoirs Chesterton tells of attending a dinner sponsored by the *Clarion* staff at the time he and the editor were fighting. G. K. found himself sitting next to a man who wanted to know if he really believed all the things he had been saying.

> I informed him with adamantine gravity that I did most definitely believe in those things I was defending against Blatchford. His cold and refined face did not move a visible muscle; and yet I knew in some fashion it had completely altered. "Oh, you *do*," he said, "I beg your pardon. Thank you. That's all I wanted to know." And he went on eating his (probably vegetarian) meal. But I was sure that for the rest of the evening, despite his calm, he felt as if he were sitting next to a fabulous griffin.

Chesterton goes on to say that one central point in his dispute with Blatchford was over free will. Blatchford was a strict determinist. Chesterton sensibly argued that if you didn't believe the will was free, it would be unreasonable to thank someone for passing you the mustard.

And yet the two men remained friends. In *Heretics* Chesterton calls Blatchford "a very able and honest journalist." Indeed, G. K. had more respect for him than Blatchford had for G. K. I have no doubt that Chesterton was thinking of Blatchford when he has MacIan call Turnbull a "good atheist . . . you clean, courteous, reverent, pious old blasphemer." It is surely a coincidence, but I discovered (with the help of my numerologist friend Dr. Irving Joshua Matrix) that if you take JT, the initials of James Turnbull, and shift each letter ahead eight steps in the alphabet (assume Z is followed by A) you arrive at RB, the initials of Robert Blatchford. Moreover, if you shift GKC back two steps you get EIA, which might stand for Evan and Ian. (And GC shifted back one step produces the FB of Father Brown.)

What finally happened to Blatchford? Fasten your seat belt! After his wife Sally died in 1921, a British medium put him in touch with her spirit and in 1923 he became a Spiritualist! "I am quite satisfied that I was with Sally," he wrote to a friend, "and I feel very happy about it." His little book *More Things in Heaven and Earth* (1926) is about his newfound faith.

I do not know whether Chesterton, who died in 1936, ever expressed relish over this conversion of his friendly enemy to miracles and a strong belief in an afterlife. It surely would have reminded him of his often-quoted aphorism, that when people stop believing in God they do not believe in nothing but in anything. Conan Doyle, too, turned to Spiritualism after abandoning Catholicism. In his history of Spiritualism, Doyle proudly lists Blatchford among prominent "materialists" who converted to spookery.

On a second level of allegory, *The Ball and the Cross* is clearly intended to mirror the conflict between Augustine's City of God, which for Chesterton was the Catholic Church, and the City of Man, which is under the control of Satan. If you are a conservative Catholic you can interpret the novel as about the conflict between the Church and Satan—a conflict destined to last until the end of the world. William F.

Buckley, in the introduction to a reprinting of his first book, *God and Man at Yale*, must have had Chesterton's novel in mind when he wrote: "The duel between Christianity and atheism is the most important in the world. I further believe that the struggle between individualism and collectivism is the same struggle reproduced on a higher level."

Readers who are non-Catholic Christians can take the novel more broadly as a conflict between Christian faith and atheism, and orthodox Jews and Muslims can similarly see the novel as a conflict between their faith and its enemies. This brings us to the third and highest level of allegory. On this level MacIan does not symbolize any particular faith, but only a faith in God. No chasm in the history of thought is deeper than that between those who believe in a God who cares for humanity and those who see no purpose or morality in the universe apart from the purposes and morals of the human species.

On this nonsectarian level, *The Ball and the Cross* is an indictment of modern culture for being essentially Laodicean. In Revelation 3:16 John attacks the church at Laodicea for being neither cold nor hot: "So then because thou art lukewarm, and neither cold nor hot, I will spue thee out of my mouth." Contrast this (as I learned from Donald E. Knuth's book *3:16*) with John 3:16, a verse that draws a sharp line between the saved and unsaved. In past centuries great thinkers were proud to let everyone know where they stood on religious questions. Today they couldn't care less. A philosopher willing to stand up and fight for either atheism or theism is considered eccentric and slightly mad. Has any philosopher of worldwide eminence since Bertrand Russell openly called himself an atheist?

Laodicean lukewarmness extends even to those who profess a religion. We still insist that our presidents call themselves Christian, but exactly what they believe is deemed irrelevant. Catholic admirers of John Kennedy seem not in the least disturbed to learn that he lost his faith early in fife. All that matters is that he went to Mass and pretended to be a Catholic. Who knows or cares what dogmas Ronald Reagan, who professes to be a born-again Protestant, really believes? Who knows what Episcopalian doctrines George Bush believes? We know he was a good pal of Billy Graham and Jerry Falwell, but were these friendships genuine, or were they carefully calculated to win political support from the ever increasing ranks of fundamentalist voters? Bill Clinton is a Baptist and Hillary is a Methodist. Does either of them believe that the tomb of Jesus became miraculously empty? A reporter is free these days to ask a politician if he or she has committed adultery, but to ask about a basic religious belief would be considered the height of bad taste.

MacIan wanted to kill a man because he made unkind remarks about Mary. Indeed, a Catholic who denies the Virgin Birth, the Immaculate Conception and the Annunciation of Mary runs the risk, if not of excommunication, of damnation. In past centuries Catholics and Protestants alike were tortured and killed for expressing skepticism about much less fundamental doctrines.

Today, you will have a difficult time discovering what any prominent Christian actually believes. Does Father Greeley, for example, who has written an article about Mary for *The New York Times Magazine*, believe in the Virgin Birth? Do such Catholic politicians as Mario Cuomo, Jerry Brown, Eugene McCarthy, Edward Kennedy, and Geraldine Ferraro? What about such Catholic writers as Pat Buchanan, Garry Wills, and Peggy Noonan? John McLaughlin, who presides over the McLaughlin Group on TV, is an ex-priest. What doctrines does he now believe? Who cares? It is not so much that the public is irreligious, but that it is lukewarm, indifferent to dogmas. It is embarrassed when views are debated in public.

Millions of Catholics and Protestants around the world now attend liberal churches where they listen to music and Laodicean sermons, and (if Protestant) sing tuneless Laodicean hymns. They may even stand and recite the

Apostles' Creed out of force of habit and not believe a word of it. If a pastor or priest dared to preach a sermon on, say, whether Jesus' corpse was actually revivified, the congregation would quickly find a way to get rid of him.

It is a scandal of American Protestantism that no one knows whether Reinhold Niebuhr did or did not believe in the afterlife taught by Jesus. I once tried to find out by writing to his widow, but she replied in a diplomatic letter that she had to let her husband's writings speak for themselves. Alas, nowhere in those writings can one find a clear answer to this question. Either Mrs. Niebuhr herself didn't know, or she wouldn't tell me. *The Ball and the Cross* is about a thick yellow fog that has settled over Christendom. Only Protestant fundamentalists and conservative Catholics, God bless 'em, are willing to take firm stands and say exactly what they believe.

It is this yellow fog that both MacIan and Turnbull detest. Along the way, in their efforts to find a spot where they can fight unmolested, they meet a Tolstoyan pacifist, a purveyor of pornography, a Nietzschean nihilist, a vague Protestant, a good citizen, a wimpish man who loves bloodshed, and a pantheist enamored of the "life force." All these men have found mushy substitutes for a robust theism or atheism. When God goes, Chesterton is telling us, half-gods arrive.

Lucifer, with his large cloven chin (the cleft symbolizes his break with God), is of course the Prince of the World. He becomes England's secret ruler, certifying as insane not only MacIan and Turnbull, but all those who have even learned of their desire to fight. Professor Lucifer is the force behind modern science, about which Chesterton unfortunately knew nothing, and especially behind what G. K. considered the false theory of evolution. It is Lucifer who invented his airship's advanced technology, and who runs the asylum in Kent where everything is handled with superb efficiency by invisible machines that keep the walls washed and sterile, and provide mushy food. The asylum, where the doctors are as crazy as the patients, is Chesterton's caricature of a secular world gone mad in its efforts to avoid coming to grips with Christianity.

One must not suppose that Chesterton favored any sort of actual warfare between believers and unbelievers. Both Turnbull and MacIan, after two remarkable dreams, are given visions of the future each has longed for. In Turnbull's dream the cross of St. Paul's is broken, and secular socialism has grown into a police state similar to the Soviet Union under Stalin or Communist China under Mao or its present leaders. In MacIan's dream the ball on St. Paul's has become invisible. The state has become a theocracy, a police state in the grip of Christian fascism. Its spirit is that of the Inquisition, the massacre of St. Bartholomew, the burning of witches, and today's Islamic nations eager for holy wars against infidels.

Note how Chesterton has anticipated the rewriting of history by the Soviet Union when he has England turn the duel into a nonevent by incarcerating everyone who knew about it. Observe how he has anticipated the Soviet Union's use of mental institutions to suppress political dissent. Both MacIan and Turnbull eventually realize that if either of their fanatical views ever captured a nation, the result would be a terrible tyranny and the disappearance of individual freedoms. Only religious tolerance and the freedoms of democracy can prevent the rise of such evil empires.

Would that Chesterton had left it at this, but instead he ends his novel with an improbable climax. At the last moment, when Turnbull witnesses a miracle, he becomes a Christian! The two swords fall to the ground in the pattern of a cross. Chesterton is of course expressing his belief that in some far-off future the entire world will become the City of God, but such an ending certainly turns off any reader who is not a conservative Catholic or a Protestant fundamentalist.

For Chesterton, the arms of the cross stretch into the universe and beyond the universe into infinite realms that transcend the world we know. The ball, in contrast, is finite and self-contained. Rationalism without faith goes round and round in circles, never escaping from the material cosmos. Perhaps H. G. Wells had the ball in mind when, in the parallel world of *Men Like Gods*, he has Jesus killed by torture on a wheel.

Other symbols in the novel are less clear. In the Bible, Michael is an archangel. But the little Bulgarian monk, one of whose names is Michael, is not an angel. I believe that Chesterton intended him to signify the Catholic Church. He is ancient, learned, and wise, but when he is persecuted and confined, he takes on the appearance of being childish:

> One talks trivially of a face like parchment, but this old man's face was so wrinkled that it was like a parchment loaded with hieroglyphics. The lines of his face were so deep and complex that one could see five or ten different faces besides the real one, as one can see them in an elaborate wall-paper. And yet while his face seemed like a scripture older than the gods, his eyes were quite bright, blue, and startled like those of a baby. They looked as if they had only an instant before been fitted into his head.

The monk's final act, when he sings, laughs, and performs a miracle, suggests how the Church, in spite of its persecutions, always bursts forth with renewed power, "reeling but erect" as Chesterton put it in *The Everlasting Man*. More than once in the novel Turnbull calls the monk a "thing," and toward the end of chapter 12 the word is capitalized. It is no coincidence that one of Chesterton's books about the Church is titled *The Thing*.

There is a mystery closely connected to the monk. In each of the cells in which the monk, MacIan, and Turnbull are confined, the walls are perfectly smooth except for one iron peg or spike. Turnbull detests it more than the "damned cocoa" he is served daily. (Chesterton hated cocoa. Here its blandness suggests the lukewarm food that the Bible says will be vomited.) So annoyed is MacIan by the spike that he rips it from the wall. "Even now I've got it out," he says, "I can't discover what it was for." And later we learn that "to this day" Turnbull does not know its purpose.

But the monk is not annoyed by the spike, or by anything else about his cell. He likes the cell's long, narrow shape, and he amuses himself by counting the squares on the floor. "But that not the best," he says. "Spike is the best . . . it sticks out." We are not told why he said this. What on earth does it mean?

John Wren-Lewis, writing on "Joy Without a Cause" (*The Chesterton Review*, February 1986), puts forth an interesting conjecture. When the monk is dropped onto the dome of St. Paul's he narrowly escapes death. After reaching the street in safety, he feels like a child again:

> Everything his eye fell on it feasted on, not aesthetically but with a plain, jolly appetite as of a boy eating buns. He relished the squareness of the houses; he liked their clean angles as if he had just cut them with a knife. The lit squares of the shop windows excited him as the young are excited by the lit stage of some promising pantomime. He happened to see in one shop which projected with a bulging bravery on to the pavement some square tins of potted meat, and it seemed like a hint of a hundred hilarious high teas in a hundred streets of the world. He was, perhaps, the happiest of all the children of men.

The monk, according to Wren-Lewis, has discovered one of Chesterton's favorite themes—that we are surrounded by "tremendous trifles," little things that ought to give us pleasure and for which we should be grateful. The monk is happy anywhere, even in prison. In his barren cell the spike breaks the smooth monotony of the walls. It is not a flower or the bird that arouses the gratitude of Byron's "Prisoner of Chillon." Still, it is something, and the monk is thankful for it, taking pleasure in its spikiness. In a later book on Robert Louis Stevenson, Chesterton wrote of Stevenson: "He loved things to stand out; we might say he loved them to stick out; as does the hilt of a sabre or the feather in a cap. He loved the pattern of crossed swords; he almost loved the pattern of the gallows because it is a clear shape like the cross." To Freudians, a spike is a phallic symbol, but Chesterton could not have had that consciously in mind.

I think Wren-Lewis is right about the spike, but it seems to me that Chesterton may also have had in mind a symbolic meaning. The blank walls of the cell represent the blank metaphysics of atheism. They are smooth and continuous like the surface of the ball on St. Paul's. But for Chesterton, history has been repeatedly broken by enormous discontinuities.

The creation of the universe was the first big bang. The fall of Satan and the fall of man were other bangs. The abrupt appearance of life, and the sudden appearance of animals with human souls, are still other discontinuities. Evolution, which Chesterton rejected without understanding, is a gradual process of change that began with the big bang. But for Chesterton, history is a series of things that "stick out." Each person's birth is a spike. Each person's rebirth is a spike. The Incarnation, when God suddenly entered the world, is a colossal peg that for Chesterton was the central turning point of human history. Each arm of the cross is a spike. And the Bible promises an apocalyptic end to history. The big fire that closes *The Ball and the Cross*, forcing Lucifer to leave the earth, is the final conflagration at the end of the world.

If there is no God to give history a transcendent purpose, then the universe and all history are ultimately pointless. But Chesterton, like any theist, believed the universe has a point as sharp as the point of a sword. "For me all good things come to a point," Chesterton writes in chapter 4 of *Orthodoxy*. Similarly, "Everything is coming to a point," MacIan says in chapter 19. Indeed, the conviction that the universe is not pointless is the main point of the novel. Perhaps that is also the basic meaning of the spike.

When we are told that "Above all, he [Turnbull] had a deep hatred, deep as the hell he did not believe in, for the objectless iron peg on the wall," perhaps Chesterton is symbolizing what in *The Everlasting Man* he calls "the halo of hatred around the Church of God." Elsewhere Chesterton liked to symbolize the Church by a key. When MacIan pulls out the spike, it proves to be a key that alters the asylum's machinery and unlocks all the cell doors.

The Ball and the Cross was first published in book form in 1909 by the John Lane Company of New York. Sheets were sent to Wells Gardner, Darton & Co., Ltd., in London, for the first British edition the following year. The novel had previously been serialized in 1905 and 1906 in *The Commonwealth*, the Christian socialist monthly mentioned earlier. It has been translated into French, Italian, Spanish, Czechoslovakian, and Polish, perhaps into other languages. Reviews of the book, almost all objecting to the plot's allegorical obscurity, appeared in British journals in 1910.*

A misprint in both American and British first editions is worth mentioning. On page 93, line 6 of these editions, "heal thy working man" should be "healthy-looking man."Also worth noting is that, on page 358, line 18, of the American edition, there is the sentence: "But God blast my soul and body!" The British edition changed this to "But Lord bless us and save us!" And on page 345 we encounter what surely is a mistake of some sort when the monk says, "I am very happy," and the narrator follows this with "said the other, alphabetically." The word "alphabetically" makes no sense. John Peterson, who edits the Midwest Chesterton News, has a plausible explanation. Chesterton liked to dictate to stenographers. Did he speak the word "affably," which was either heard as "alphabetically" by the stenographer or later so misinterpreted from her shorthand?

---

*Here is a partial list, with volume and page numbers: *Atheneum* 1:337; *Catholic World* 91:836; *Independent* 68:417; *Nation* 89:651; *Outlook* 94:504; *Saturday Review* 109:337; *Spectator* 105:285.

In later years Chesterton expressed dissatisfaction with the novel. Here is a comment from his autobiography:

> I think the story called *The Ball and the Cross* had quite a good plot, about two men perpetually prevented by the police from fighting a duel about the collision of blasphemy and worship, or what all respectable people would call, "a mere difference about religion." I believe that the suggestion that the modern world is organized in relation to the most obvious and urgent of all questions, not so much to answer it wrongly, as to prevent it being answered at all, is a social suggestion that really has a great deal in it, but I am much more doubtful about whether I got a great deal out of it, even in comparison with what could be got out of it.

In a copy of *The Ball and the Cross* owned by Father John O'Connor, the original of Father Brown, Chesterton inscribed a poem that begins:

> This is a book I do not like,
> Take it away to Heckmondwike,
> A lurid exile, lost and sad
> To punish it for being bad.
> You need not take it from the shelf
> (I tried to read it once myself.
> The speeches jerk, the chapters sprawl,
> The story makes no sense at all)
> Hide it your Yorkshire moors among
> Where no man speaks the English tongue.

Maurice Evans, in a little book of praise for Chesterton, called *The Ball and the Cross* a "confused allegory." Although its central symbolism is obvious, the role of the monk is certainly obscure and unsatisfactory. At the beginning, when he debates Lucifer in the airship, he is a man of vast learning and acute intelligence. Yet at the end he is a helpless, bizarre idiot. Chesterton was probably trying to say that this is how the church appears when pushed down by a secular culture, but even so it is hard to believe that any of Chesterton's readers, or G. K. himself when he was older, could consider his monk a good symbol of the Church.

A philosophical but non-Christian theist such as myself can enjoy the novel on what I have called its third level of allegory. In a sense I admire a Protestant fundamentalist or a conservative Catholic more than someone who either has no religious opinions or, if he or she has, is too timid to talk about them. If you are an atheist or agnostic of the sort that has no interest whatever in the clash between theism and atheism, you will likely find the novel not worth your time. On the other hand, you may still enjoy reading it for its colorful style, with its constant alliteration, amusing puns, and clever paradoxes; for its purple passages about sunsets, dawns, and silver moonlight; and for the humor and melodrama of its crazy plot.

POSTSCRIPT

My speculations about the meaning of the spike brought several letters. Reverend Charles B. Gordon suggested that Satan had put it in the cells as a convenient means by which prisoners could hang themselves. He elaborated on this conjecture in a note published in *The Chesterton Review* (August 1997). I quote his final paragraph:

> I have argued that the spikes in the cell walls are intended to serve as gallows or gibbets. For this reason the spike in his cell reminds Michael of another instrument of execution: The cross upon which Jesus died. The plausibility of this identification is increased if the reader associates the three spikes with the spikes or nails employed in the Crucifixion. It is not surprising that a figure representing the Church should have a particular affection for an object he finds reminiscent of the cross of Jesus. Once it is recognized that Chesterton intends an association between the spikes and the Cross, it is

easy to understand why, when MacIan wrenches the spike from the wall of his cell, the doors to the cells are thrown open. The Cross is thereby allegorically identified as the key to the release of humankind from captivity.

Dale Ahlquist found the following statement in Chesterton's *Illustrated London News* column (October 30, 1926): "I think I shall try some day to write a huge philosophical and critical work called *The Point: Its Position, Importance, Interest, and Place in our Life and Letters*."

I located several relevant passages in Chesterton's *Orthodoxy*. In chapter 4: "For me all good things come to a point, swords for instance." In chapter 5 the point becomes a spike:

> And then followed an experience impossible to describe. It was as if I had been blundering about since my birth with two huge and unmanageable machines, of different shapes and without apparent connection—the world and the Christian tradition. I had found this hole in the world: the fact that one must somehow find a way of loving the world without trusting it; somehow one must love the world without being worldly. I found this projecting feature of Christian theology, like a sort of hard spike, the dogmatic insistence that God was personal, and had made the world separate from Himself. The spike of dogma fitted exactly into the hole in the world—it had evidently been meant to go there—and then strange things began to happen. When once these two parts of the two machines had come together, one after another, all the other parts fitted and fell in with an eerie exactitude. I could hear bolt after bolt over all the machinery falling into its place with a kind of click of relief.

I mentioned in my introduction that *The Ball and the Cross* had been serialized in *Commonwealth* in 1904–1905. For some reason, unknown to me, only the first half was published. Chapter 9 ended with "to be continued" but the remaining eleven chapters never appeared.

Several articles on *The Ball and the Cross* have appeared in *The Chesterton Review*. Especially notable is John Coates's essay, "The Ball and the Cross and the Edwardian Novel of Ideas" (February 1992).

On the question of whether the universe has a point, the physicist Steven Weinberg, in his 1977 book, *The First Three Minutes*, makes a statement that has often been quoted. After describing how beautiful the earth is, yet how hostile the universe, he adds: "The more the universe seems comprehensible, the more it also seems pointless."

INTRODUCTION TO
THE NAPOLEON
OF NOTTING HILL

W hen *The Napoleon of Notting Hill* was first published in 1904, by the London house of John Lane: The Bodley Head, Gilbert Keith Chesterton was thirty. His reputation as a writer was just getting under way. He had earlier published two small books of poetry, two essay collections, and biographies of the painter G. F. Watts and of Robert Browning. England was slowly becoming aware of a strong, unusual new voice in letters, but it was G. K.'s first novel that permanently established his long and productive career.

The Napoleon of Notting Hill was such a strange sort of fantasy that critics did not quite know what to make of it, though for the most part they enjoyed and praised it. Like the popular science fiction of H. G. Wells and others, it was set in the future, its curtain going up in 1984, eighty years after the book's appearance. Was this the reason George Orwell

This introduction first appeared in G. K. Chesterton, *The Napoleon of Notting Hill* (New York: Dover, 1991).

chose the same year for his antiutopian novel? No one seems to know. Orwell read and admired G. K., but his choice of 1984 may have been no more than a curious coincidence.

The two visions of the future, Orwell's and Chesterton's, have little in common with each other or with the world's shape today. In Orwell's novel, technology has advanced as rapidly as Wells and Jules Verne knew it would. Chesterton, whose interest in science was minimal, made no effort to project technological trends. Indeed, in his first chapter he heaps scorn on "wiseacre" predictions that vehicles will go faster and faster, perhaps whirl around the earth so rapidly that a person on board could carry on a conversation with someone on land. Later in the novel, Adam Wayne ridicules the notion of "tramcars" going to the moon. Chesterton's London, in the last two decades of this century, looks exactly like the Edwardian London of 1904. No motorcars crowd the streets. No airplanes cross the skies. Londoners still move about mainly in horse-drawn hansom cabs, although horseless double-deck omnibuses (first licensed in London in 1904) also roam the streets.

"There is no such thing . . . as night in London," we are told, but the illumination in 1984 is still from gas lamps. There are no telephones. In short, London remains unelectrified. In the fierce battles that take place on Notting Hill, not a single shot is fired. The fighting is all hand-to-hand, with swords and battle-axes. Humanity, playing the game that G. K. calls "Cheat the Prophet," has listened to all the predictions by the wise men, then cheated them by going on for eighty years with no technological change. Needless to say, history has cheated the prophet Chesterton.

In G. K.'s defense, however, it must be said that it was never his intention to imitate Wells by trying to guess the future, but only to use London as the setting for a humorous fantasy that would, like all his later fantasies, convey philosophical messages. At the end, his story becomes a theological allegory, but on a lower level its message is political. Chesterton saw the world of 1904 drifting toward larger and larger empires as big nations swallowed little nations, with an accompanying decline of local folkways and patriotic pride. The last of the small nations to go, by 1984, is Nicaragua. Early in the novel there is a startling scene in which Nicaragua's deposed president, Juan del Fuego, stabs his own palm so that his blood will stain a yellow piece of paper, honoring the red and yellow colors of Nicaragua's flag. (Chesterton has been cheated again. Today, Nicaragua remains independent, and its flag is blue and white.)

By 1984, world peace has been achieved, and England is so efficiently governed, and its citizens so obedient, that it has almost no soldiers or police. Democracy has died because everyone is content to let the ruling class rule. It is the end in England of all political ideologies. Interest in revolution has been replaced by a resigned acceptance of slow, safe evolution. There is no longer any "reason for any man doing anything but the thing he had done the day before." So efficient is the bureaucracy that a pall of dullness has settled over the land. Life is drab, colorless, and humdrum, as unromantic as the Soviet empire depicted in Orwell's fantasy novel.

England has abandoned its hereditary monarchy. There is still a king, but he is chosen like a juryman from a rotation list of civil servants. Queens, apparently, are no longer possible. Indeed, London women have become so invisible that there is not a single woman in the novel—not the least of the curiosities of this curious book. Within limits imposed by the wealthy ruling class, the king is free to do whatever he likes. If he does something idiotic, nobody cares so long as the nation's monotonous stability is not seriously disturbed. One thinks of public indifference to the foolish behavior of our own political leaders. Harry S. Truman threatens to kick a music critic in the groin because he has criticized the president's daughter's singing. John F. Kennedy sleeps with the girlfriend of a notorious gangster, and passes Marilyn Monroe over to his brother

Bobby. Jimmy Carter collapses when he tries to win a marathon race. Ronald Reagan sends the Ayatollah a cake and an inscribed copy of the Bible. (Perhaps Nancy Reagan's astrologer advised it.) The public may find such acts amusing, but does it really care? Are we getting closer to Chesterton's own vision of 1984 in which few bother to vote, and nobody much minds how idiotically political leaders behave?

Things start to change in London when the crown lands on the head of a lowly bureaucrat named Auberon Quin. Oberon was, of course, Shakespeare's king of the fairies, and Chesterton does not hide the fact that his King's name suggests the "midsummer madness" that pervades the book's dreamlike narrative. Although the King has a flair for writing and painting (he publishes poetry under the *nom de plume* of Daisy Daydream, and reviews books as Thunderbolt), he is a total cynic who thinks everything is funny, from iron railings to cornfields, sunsets, and even the stars. He is an art-for-art's-sake aesthete interested only in amusing himself, like the small child he resembles in his short stature and round baby face. A man may strike a lyre, he says, and recite with Longfellow that " 'Life is real, life is earnest,' " then he goes and "stuffs alien substances into a hole in his head." Like the kangaroo, a human being is one of nature's farces.

So boring is life that the King longs for miracles to break the monotony, such as a lamppost turning into an elephant. Because such events never happen, he does his best to enliven things by playing the part of the court jester. He tells nonsensical jokes. He loves inflicting pranks on humorless associates, such as sitting on their hats. After his coronation, the King's first public act is to put his coat on backward. King Auberon is like the hero of Rafael Sabatini's novel *Scaramouche* (1921), who "was born with a gift of laughter and a sense that the world was mad."

Mingling with his people one afternoon, the King comes upon a red-haired child, in the suburb of Notting Hill, who is wearing a paper military hat and brandishing a wooden sword. While the King is praising the lad for his patriotism, a wild idea enters his head. Why not add color and romance to dull London by reviving local patriotism? Each suburb will have walls around it, with a gate to be closed at sunset and guarded by soldiers in brightly colored uniforms. Each suburb will be given its own coat of arms, its tocsins, its flag, its legends, its patriotic songs. In brief, why not revive the medieval city-states with all their color, pomp and rich symbolism?

Vastly amused by this scheme, the King issues a Charter of the Free Cities. Five boroughs—North Kensington, South Kensington, West Kensington, Bayswater, and Notting Hill—are each to be ruled by a Lord High Provost, dressed in robes that display the city's official color. The five colors—red, yellow, blue, green, and purple—are the same as the colors of the five regions of Oz. A red lion is made the symbol of Notting Hill.

No one in England takes the King's joke seriously. After all, they know him to be an irrelevant fool. Ten years later, Auberon is still amused, but now the plot takes an ominous turn. Plans are under way to run a large thoroughfare through Notting Hill that will demolish Pump Street, a small street of shops near the center of Notting Hill (like the heart, which pumps blood, near the center of the human body). The provost of Notting Hill is none other than the little boy with red hair, now nineteen, grown tall, with a large nose and baby-blue eyes that are described as bold, uncanny, queer, and innocent. He had been so inspired by the King's paean to patriotism that when the Charter of the Free Cities was proclaimed, he alone in England took the joke seriously. Under no circumstances will he allow the proposed road to destroy Pump Street. If necessary, he will wage war to defend it. King Auberon collapses with laughter when he hears this, thinking the youth is a kindred soul going along with the jest. But no, the provost of Notting Hill is in deadly earnest. Indeed, he totally lacks a sense of humor.

The young man's name, Adam Wayne, suggests the innocence of Adam before the Fall. (At the end of the novel Wayne
uproots a giant oak tree, a symbol of the tree of knowledge in
the Garden of Eden, during a battle in Kensington Gardens.)
Wayne recruits an army to defend his free city, and a war
results which the King enjoys immensely. He appoints himself special correspondent for a London newspaper, writing
long, lurid dispatches from the front, under the by-line of
Pinker, that parody the florid journalism of Chesterton's own
time. He even calls for the King's resignation!

I will not spoil the reader's fun by revealing any more of
this preposterous plot, except to say that the battles for
Notting Hill are hugely comic. Although thousands are
slain, and blood crawls through the streets like "great red
serpents, that . . . shine in the moon," the scenes have as
little relation to modern warfare as battles in old chivalric
romances, or wars fought with cardboard soldiers in the
toy theaters which Chesterton loved.

There are two accounts, both by G. K., of how he came
to write his novel. Here is what he said to his friend W. R.
Titterton, who recorded the words in his little book *G. K.
Chesterton: A Portrait* (London: Douglas Organ, 1936):

> "It came to me in a flash when I was walking down a cer-
> tain street in Notting Hill. There was a row of shops. At
> one end was a public-house; somewhere at the farther
> end I rather think there was a church. And on the way
> there was a grocer's, there was a secondhand bookshop,
> there was an old curiosity shop where they sold, among
> other things, arms. There were, in fact, shops supplying
> all the spiritual and bodily needs of man.
>
> "And all at once I realized how completely lost this bit
> of Notting Hill was in the modern world. It was asked to
> be interested in the endowment of a public library in
> Kamschatka by an American millionaire, or a war
> between an oil trust and another oil trust in Papua, or the
> splendid merger of all the grocery interests in Europe and
> America or the struggles between the brewers and the
> Prohibitionists to give us worse beer or less beer.

"In all these world-shaking events this little bit of Notting Hill was of no account. And that seemed idiotic. For to this bit of Notting Hill the bit was of supreme importance.

"In the same instant I saw that my Progressive friends were more bent than any on destroying Notting Hill. Shaw and Wells and the rest of them were interested only in world-shaking and world-making events. When they said, 'Every day in every way better and better,' they meant every day bigger and bigger—in every way.

"Now in Nature there is nothing like this continuous expansion—except growth, which ends when the creature reaches maturity. And though, somewhat unjustly as I think, I may be cited as a case to the contrary, the creature does not expand equally in all directions.

"I saw that these Progressives were obsessed with the idea of dilation. There is such a thing as a dilated heart, which I am told is a disease. There is such a thing as a dilated, or swelled, head. But the typical case of a creature who dilated equally all round is that of the imperially minded frog who wanted to be a bull, and dilated until he burst. . . .

"In that half-second of time, gazing with rapt admiration at the row of little shops, nobly flanked by a small pub and a small church, I discovered that not only was I against the plutocrats, I was against the idealists. In the comparatively crystalline air of that romantic village I heard the clear call of a trumpet. And, once for all, I drew my sword—purchased in the old curiosity shop—in defense of Notting Hill."

This is how Chesterton told it in his posthumously published autobiography:

I was one day wandering about the streets in that part of North Kensington, telling myself stories of feudal sallies and sieges, in the manner of Walter Scott, and vaguely trying to apply them to the wilderness of bricks and mortar around me. I felt that London was already too large and loose a thing to be a city in the sense of a citadel. It seemed to me even larger and looser than the British Empire. And something irrationally arrested and pleased my eye about the look of one small block of little lighted shops, and I amused myself with the supposition that these alone were to be preserved and defended, like a hamlet in a desert. I found it quite exciting to count them and perceive that they contained the essentials of a civilization, a chemist's shop, a bookshop, a provision merchant for food, and public-house for drink. Lastly, to my great delight, there was also an old curiosity shop bristling with swords and halberds; manifestly intended to arm the guard that was to fight for the sacred street. I wondered vaguely what they would attack or whither they would advance. And looking up, I saw grey with distance but still seemingly immense in altitude, the tower of the Waterworks close to the street where I was born. It suddenly occurred to me that capturing the Waterworks might really mean the military stroke of flooding the valley; and with that torrent and cataract of visionary waters, the first fantastic notion of a tale called *The Napoleon of Notting Hill* rushed over my mind.

The inept working title for the novel was *The Devil Amongst the Cattle*. Later, Chesterton suggested *The Lion of Notting Hill* and *The King and the Madman*. Apparently it was his editor who came up with the final title. The book sold well, and by 1949 was in its eighth printing. It was translated into French in 1912, and into Spanish in 1941. A Paulist Press paperback, with an introduction by Father Andrew Greeley, was on the United States market for a short time starting in 1978.

A play version had a one-night performance on March 7, 1926, at London's Regent Theatre. Little is known about it. In April 1984, a musical version, produced in London by Canadian promoters, ran for six nights at the Old Vic Youth Theatre, and later had seven performances at the Jeannetta Cochrane Theatre. Rodney Archer and Powell Jones wrote the script, with music by Chuck Mallet and lyrics by David Head. Ironically, although there are no female characters in the novel, most of the actors were women. Extreme liberties were taken with Chesterton's plot. A Pump Street butcher named Wayne, after having four daughters, insists on naming his fifth daughter Adam. She strides into King Auberon's chamber to sing, "I am the child of your charter. . . ." Juan del Fuego Lovemore, president of Cornucopia, descends onto the stage in a big red and yellow balloon.*

Many episodes in *The Napoleon of Notting Hill* foreshadow episodes in Chesterton's later books The King is amused by the way men in frock coats, seen from behind, look like dragons walking backward The character named Sunday is similarly described in *The Man Who Was Thursday.* Auberon likes to stand on his head. So does Gabriel Gale the mystery-solver in *The Poet and the Lunatics.* The upside-down crucifixion of Saint Peter, which Wayne considers an obscene joke, is described by Gale as a time when Peter "saw the landscape as it really is: with the stars like flowers, and the clouds like hills, and all men hanging on the mercy of God."

King Auberon invents a dancing language. So does Professor Chadd in *The Club of Queer Trades.* In several later books Chesterton repeats the whimsy—from Wayne's forgotten book of poetry, *Hymns on the Hill*—of describing nature with the metaphors of a city: wind whips around a street corner like a hansom cab. Swords play a prominent role in many of G. K.'s books and poems. He even carried a cane with a handle that pulled out to become a rapier in case, he once explained, he ever had to defend a lady in distress.

The name of Turnbull, the quiet owner of the curiosity shop on Pump Street who turns into Wayne's bull of a general, is used again for the atheist in *The Ball and the Cross.* Even the Great Battle of the Lamps, which plunges London into pitch darkness when Wayne cuts off the gas,

---

*My information about this musical relies entirely on a letter from Dr. William E. Griffiths, of London, that appeared in *The Chesterton Review* (February 1985).

is echoed in the memorable introduction to *Heretics* where Chesterton praises rational argument. When the lights of reason go out, he warns, men indeed seem to go blind. "And there is war in the night, no man knowing whom he strikes. . . . Only what we might have discussed under the gas-lamp, we now must discuss in the dark."

When *Napoleon* came out in 1904, critics recognized at once that the illustrations by William Graham Robertson— seven tipped-in pictures, along with a map of Notting Hill— gave King Auberon the unmistakable small size and round face of Max Beerbohm. In his own autobiography *Time Was* (1931), Robertson makes it clear that this caricature of Beerbohm was intended:

> Once, when I was particularly busy, there arrived a bulky parcel containing a novel by G. K. Chesterton, which I regarded dubiously, wondering how I should find time to read it. But, as I glanced at it, my eye fell upon the first phrase—"The Human Race, to which so many of my readers belong—" and at once I dispatched to John Lane, the publisher, an enthusiastic recommendation. A book which began like that must be all right; no one could afford to throw away such a gem in the opening sentence who had not plenty more to follow.
>
> I was enlisted to illustrate the work which turned out to be a witty and fantastic picture of a future England, reigned over by an elected King who appeared to be none other than Max Beerbohm, or at least a recognizable caricature of him. But Max, himself a caricaturist, was fair game; besides, he was in the secret and made no objections, and John Lane used to ask him and me to meet Mr. Chesterton, so that the collaboration between novelist, model, and illustrator might become the more harmonious.

Notting Hill, a section of Kensington northwest of Hyde Park, is mentioned in three Sherlock Holmes mysteries. A man named Selden, in *The Hound of the Baskervilles*, is called "the Notting Hill murderer." Louis La Rothière, in

"The Bruce Partington Plans," lives in Notting Hill. And in "The Red Circle," Inspector Gregson treats a woman "with as little sentiment as if she were a Notting Hill hooligan."

A legend has it that Kensington was named after the fairy Kenna, daughter of King Oberon. The British novelist Roy Kerridge, in an article on "The Prophets of Notting Hill" (*The Chesterton Review*, fall/winter 1979–80), gives a frightening description of what Kensington's Notting Hill district was like in 1979. He calls it "one of the most squalid, romantic, exciting, and lawless parts of London." No longer were its streets clean and quiet as they had been in 1904. The area had become a slum for poor Irish, blacks, West Indians and young counterculture whites. There was heavy drinking, drug taking, prostitution, porno movie houses, and racial violence incited by gangs first called Teddy Boys and then called Skinheads. Since 1980, however, Notting Hill has undergone considerable renovation. Kerridge tells of an underground station wall sprayed with leftist graffiti. "We are all insane," a slogan reads. "There is no such thing as sanity." Underneath someone has added: "Right on! I never use the word 'insane' anymore." In response to this a third person has scribbled: "Me neither. I say 'Gone crackers.' " A final comment: "Yeah! Gone crackers is the one! Walk cool, talk cool, think cool. Be the first in your neighborhood to have gone crackers!"

Chesterton wanted the Hill to return to a romantic past. "So do I," Kerridge concludes, "but not to a past of riots and bloodshed." He refers to the "Second Prophet of Notting Hill," the late Colin MacInnes (he died in 1975), whose 1959 novel *Absolute Beginners* paints a horrendous picture of Notting Hill's deterioration during the 1950s when young whites in England became hooked on drugs, jazz, rock, free sex, motorcycles, and violence. Like their counterparts here in the sixties, the youths in the novel think of themselves as "absolute beginners," the vanguard of a brave new world, even though they haven't the foggiest notion of how to bring it about or how to run it if it should materialize.

Absolute Beginners is narrated by an unnamed eighteen-year-old who supports himself as a freelance porno photographer, and who writes about his underworld, using hundreds of annoying slang terms that MacInnes later confessed were "almost entirely invented." Blacks are called spades; indeed, an earlier MacInnes novel about London is titled *City of Spades*. MacInnes was born in South Kensington, not too far from the house in Sheffield Terrace, North Kensington, where Chesterton was born. *Absolute Beginners* culminates with the terrible race riots that broke out in Notting Hill in 1958. Notting Hill, by the way, is called Napoli, I assume in honor of G. K.'s fantasy novel.

The political message of Chesterton's novel is that a drift toward empires is evil; that humanity should reverse this trend by allowing small nations to remain independent, free to preserve their patriotism and their folkways. During the Boer War, England's last great imperialist victory, Chesterton was one of a small but highly vocal minority of British writers (Beerbohm and Hilaire Belloc were others) who from the start vehemently opposed England's role. Bertrand Russell at first defended England, then later changed his mind and became pro-Boer.

Curiously, George Bernard Shaw, Conan Doyle, and H. G. Wells strongly defended England's part in the South African war. "Indeed," writes Chesterton in his autobiography, "he [Wells] defends the only sort of war I thoroughly despise, the bullying of small states for their oil or gold; and he despises the only sort of war that I really defend, a war of civilizations and religions to determine the moral destiny of mankind." In *Crimes of England* Chesterton calls the South African war "a dirty work which we did under the whips of money-lenders."

Chesterton did not sympathize with the Boers for pacifist reasons—he was never a pacifist—nor did he defend the Boers (religious farmers of Dutch descent) for Marxist reasons. He was against the war simply because he thought the Boers were right in defending their Orange Free State and their Transvaal Republic, with its rich gold mines and cheap labor. I wonder if it ever occurred to G. K. that the Dutch had as little right as the British to rule the native blacks?

It was Chesterton's admiration for local patriotism, for the small against the big, that is the most prophetic aspect of his fantasy. The years between 1904 and 1984 were years in which trends moved both ways. As the Soviets enlarged their empire, the British and French empires began to unravel. Chesterton would have applauded the latter as much as he would have deplored the former. Today, at the start of the 1990s, history keeps winning the game of "Cheat the Prophet" as we watch the breakup of the Soviet empire and the death throes of communism. Europe is now aflame with Wayne-like patriotism, although today it is called nationalism. To everyone's amazement, the people in Soviet-dominated nations, and even in Moscow, turned out not to have been permanently mesmerized by Marxist rhetoric. Gorbachev is Russia's King Auberon. As soon as he, in a fit of midsummer madness, told the captive countries they could do as they pleased, a million Adam Waynes leaped up to brandish their swords, demanding free elections and free markets. The despotic Ceausescus, "king" and "queen" of Romania, were executed like ancient tyrants. It remains to be seen if the leaders of these fledgling democracies will preserve their nations' newfound freedoms or, like Adam Wayne, become Napoleons corrupted by their own powers.

In the United States, as in other industrialized countries, the battle between bigness and smallness continues, even though it is fought with words and laws rather than with swords or bullets. *Small is Beautiful: Economics As If People Mattered*, by E. F. Schumacher, was a best-seller in 1975. Mergers and takeovers are still with us, but Ma Bell has been broken up into Baby Bells, and the largest firm peddling the junk bonds that supported mergers has gone bankrupt. Norman Mailer, when he ran for mayor of New York City many years ago, proposed that the city become a separate state, presumably with himself as its Napoleon. Quebec longs to break loose from Canada. Staten Island now wants to be a separate city, its secession supported by state senator John Marchi. The black Roxbury section of Boston is fired with enthusiasm to become the independent city of Mandela.

It is not surprising to learn that *The Napoleon of Notting Hill* was a favorite of the Irish patriot Michael Collins, and I would guess it is still admired by today's Irish rebels. And yet, America continues to grow more homogeneous. Local dialects are fading, streets are becoming more alike with the same neon signs, fast-food chains, and giant malls. The populace seems to care less and less about economic crises and political ideologies as it sits entranced by the sex, violence, and car chases on movie and television screens. Why worry about the national debt when you can enjoy the latest news about Jim and Tammy Bakker, or Donald and Ivana Trump?

On a deeper level, *Napoleon* concerns the eternal battle between the cynic, for whom everything is a joke, and the fanatic whose passionate faith is untempered by skepticism or humor. It is not until the novel's last chapter that this level of allegory becomes explicit. The humorless fanatic Wayne and the cynical King, both fallen in battle, speak to one another in the darkness.

The King taunts Wayne with the notion that the entire universe could be a monstrous joke, created by an idle God for his own amusement. The stars are his "idiot fireworks." The sun and moon are the "eyes of one vast and sneering giant," alternately winking in jest. We are beasts made funnier by walking on our hind legs.

Even if this is so, Wayne answers, we should not curse God but thank him for the jest. If we take the joke seriously and enjoy it, are we not winners of the game?

Dawn breaks, symbolizing the light that will illuminate the book's deeper meanings. "Wayne," the King says, "it was all a joke." Although this revelation astonishes Wayne, he does not waver in his faith. He thanks the King for the unintended good his joke has produced. We are both mad, he says, because we are the two sides of a brain that has been bisected. "In healthy people there is no war between us. We are but the two lobes of the brain of a ploughman." The antagonists walk off together like Don Quixote, the humorless man of faith, and his skeptical, comical squire Sancho Panza, into an unknown adventure. Like the ending of the film *Casablanca*, it is the beginning of a beautiful friendship.

For Chesterton, this union of fanatical love and cynical laughter is best exemplified in Christianity, its medieval cathedrals swarming with grotesque gargoyles. Those of us who are philosophical theists, even those who are atheists, can accept the book's theme in a broader sense. Unless the love of humanity is moderated by humor and doubt, it can plunge the world into tragedy. Savonarola, Hitler, Stalin, and Mao were humorless men who thought they loved humanity when they loved only an ideology. Who can imagine Hitler laughing at Charlie Chaplin's *The Great Dictator*, or Stalin chuckling over scenes in Orwell's *1984*? The crusaders, the inquisitors, and the witch burners all thought they loved humanity.

There is a dreadful passage in Chesterton's novel in which Wayne deems religious wars the only wars worth fighting for because they are waged for the "souls of men," and not for such "tomfooleries" as money, land, and natural resources. I was tempted to put this down to one of Chesterton's blind spots until I realized it was not the author speaking but his fanatical Adam Wayne. At least, I hope so. The great religious wars of the past, the religious wars going on today, are surely among the most foolish in history. It is bad enough to see all history as a joke. It is even worse to see history as progress toward a religious goal that justifies the murder of heathens and heretics.

On the deepest level of all—and this is meaningful only to theists—is the union of love and laughter that

Chesterton believed was characteristic even of God. G. K. liked to write about bizarre aspects of nature, like kangaroos and pelicans, as God's little jokes, expressing a transcendent laughter too loud for us to hear. In *Orthodoxy*, his greatest work of nonsectarian Christian apologetics (he wrote it long before his conversion to Catholicism), he closes by reminding us that Jesus concealed neither his tears nor his anger:

> Yet He restrained something. I say it with reverence; there was in that shattering personality a thread that must be called shyness. There was something that He hid from all men when He went up a mountain to pray. There was something that He covered constantly by abrupt silence or impetuous isolation. There was some one thing that was too great for God to show us when He walked upon our earth; and I have sometimes fancied that it was His mirth.

POSTSCRIPT

The following note appeared in *Midwest Chesterton News* (May 10, 1991):

Chesterton's Pump Street Hoax

I thought it would be interesting if in my introduction to *Napoleon* I could reproduce a photograph showing what Pump Street, the center of the novel's action, looks like today. A London friend, mathematician David Singmaster, volunteered to visit Notting Hill and take some snapshots of Pump Street. I heard from him too late to include the results of his research in the Dover book, so let me recount them here for readers of *MCN*.

It turns out that there *is* no Pump Street in Notting Hill, nor was there ever such a street. I knew that Chesterton often made up London street names—in the Father Brown tales, for example—but it was a map in the first edition of *The Napoleon of Notting Hill* which flim-flammed me. (Londoners, of course, were not fooled for a moment.) It clearly shows a short street in Notting Hill, running diagonally to connect Pembridge Square with Clanricarde Gardens. The illustration reproduces Chesterton's map, the arrows indicating the imaginary Pump Street.

All streets in the area close to where Chesterton put his mythical street are now residential, with apartment buildings typical of mid-Victorian times. The only commercial street nearby is a short stretch along Moscow Road between Pembridge Square and Ossington Street. It contains a laundromat, a pub, and a small grocery store.

We know from Chesterton's autobiography (I quote the relevant passage in my introduction) and elsewhere that he did indeed visit the area of Notting Hill where he found a row of shops which figure so prominently in the novel. It would be interesting to know on just what street those shops once flourished, perhaps even to discover an old photograph of them, now that we know they were not on a street called Pump!

Bernard Bergonzi's informative introduction to an Oxford Press reprint of The Napoleon of Notting Hill (1994) can be found in *The Chesterton Review* (November 1993). See also, in the same journal (November 1999), "The Meanings of *The Napoleon of Notting Hill*, by Australian political scientist and science-fiction writer Hal G. P. Colenatch.

Notting Hill, a comedy starring Hugh Grant and Julia Roberts, was released in 1999. An Associated Press story described the film as "a valentine to one of London's trendiest, most colorful neighborhoods, an alternatively stylish and rundown area, chockablock with hip restaurants and funky shops, secret gardens and seriously grand homes." No reviews of the film mentioned Chesterton's novel.

THE FAITH OF
WILLIAM BUCKLEY

'␣'ve often wondered: If, on the unlikely chance I had the privilege of interviewing William Buckley, what questions would I ask? I decided I would ask not a single question about his political or economic views because I would know in advance how he would reply. For decades Buckley has made his opinions on such topics abundantly clear. The questions I would ask would be about his faith.

We live at a time when most persons, famous or otherwise, are extremely shy about revealing their religious beliefs. It has long been true that if someone tells you he or she is a Protestant, it tells you nothing about what that person believes. In recent decades this has become increasingly true of Catholics. There may be unity in the Vatican, but among the church's theologians and priests,

This review first appeared in the *Los Angeles Times Book Review*, April 12, 1998.

as well as among laymen around the world, there is a widening spectrum of conflicting opinions.

On the right of the continuum are the orthodox. They take the Bible to be the inerrant word of God when properly interpreted. They believe their church is the one true faith, its popes tracing back to Peter. They are certain that the great miracles of both testaments took place as described, and that miracles continue to occur, though, for reasons known only to God, less frequently. They believe that God continually reveals new truths to his church, and that popes are infallible when they announce new doctrines.

On the far left of the spectrum are the liberals. The most radical see the Bible as swarming with historical errors. They deny papal infallibility, the bodily resurrection of Jesus, the virgin birth, and other doctrines that have long been central to Catholic faith. On most religious questions they have little disagreement with liberal Protestants and Reform Jews. One wonders why they choose to remain Catholics. Their response is that the church they love is capable of slow change and they wish to remain inside to help change it.

In between the extreme ends of the spectrum are moderates with all shades of beliefs. In past ages the congregations of Christian churches, Catholic and Protestant alike, held a surprising unity of opinions. Today this applies only to fundamentalist or evangelical Protestant churches, and to such fringe sects as Mormons, Seventh-day Adventists, and Jehovah's Witnesses. A typical mainline Protestant church today is like a box of tacks pointing in all directions. This is slowly becoming true of Catholic congregations. Many attend mass out of force of habit. Having been raised in their faith they find it comforting to practice the old familiar rituals, smell the incense, enjoy the music, and even to recite creeds they no longer take seriously.

Everywhere on the liberal Christian front there is this infuriating vagueness. Consider, for example, the Christian belief in the bodily resurrection of Jesus. Not his appearance in visions, but the actual vanishing of his corpse from the tomb—a risen Christ so real that doubting Thomas could put a finger into the holes on the Lord's palms. We know that Billy Graham and Jerry Falwell believe this. Did Norman Vincent Peale? Does Robert Schuller? Do Catholics Father Andrew Greeley and Garry Wills?

Life beyond the grave was at the heart of Jesus' teaching. Did Reinhold Niebuhr, our nation's great Protestant theologian, share this belief in an afterlife? Amazingly, nobody knows. I once tried to find out by writing to his widow. She refused to tell me. Perhaps even she didn't know. She advised me to read her husband's books and let them speak for themselves. Alas, they give no hint of what he believed about immortality.

A similar fog saturates liberal Catholic opinions. Hans Küng, the influential Swiss theologian, holds views that differ in no essential way from those of liberal Protestants. His good friend Father Greeley, the University of Chicago's sociologist and popular Catholic author, is even harder to pin down. He once wrote an article about Mary for *The New York Times Magazine* (December 15, 1974). It was headed "Hail Mary: A revival of devoted interest in the only religious symbol asserting the femininity of the Ultimate." Nowhere in this hymn to Mary did Greeley disclose his opinions on the virgin birth, the immaculate conception, Mary's perpetual virginity, or her assumption into heaven.

Father Frederick Copleston was the Jesuit author of a splendid multivolumed history of philosophy as well as a book about Saint Thomas Aquinas and other works reflecting his early Thomism. In later years he turned away from Aquinas toward an Hegelian approach to philosophy. When he died in 1994, no one had the slightest notion of what he believed about any major dogma of his faith.

In light of all this doctrinal fuzziness, I have wondered for decades exactly where Buckley stood. I always read him with pleasure, even when I disagreed with him, because he writes so provocatively and so well. When I saw on sale his thirty-ninth book, *Nearer, My God* (Doubleday, 1997), I bought it at once. Would I find out at long last, I hoped, exactly what sort of Catholic he was and is?

I was not disappointed. We know now, as only Buckley's wife, relatives, and best friends have known, that Buckley is not only orthodox. He is ultraorthodox.

Because I am interested mainly in Buckley's religious views I will not be concerned here with biographical facts. Born into a wealthy and devout Catholic family, the sixth of ten children, his book is dedicated to his mother whom he considers a saint. For decades, he tells us, he delayed writing a confessional, but finally decided to undertake it. Why? He answers with three lines of an unintended limerick.

> The reason for this
> You can probably guess:
> I felt I owed something to God.

Buckley's only child, Christopher, was raised a Catholic, but Pat, his Episcopalian wife, has not converted. Buckley writes that he has never "exerted pressure" on her to join his church. Indeed, even while conversing with non-Catholic friends, or when writing his newspaper columns or his many earlier books, he has considered it bad taste to talk about his religion.[1] "I am not remotely qualified as a theologian or historian of Christianity." He comes through his book as a person brought up a Catholic who has never doubted the church's essential doctrines. "My faith has not wavered," he states at the end of his introduction.

Buckley retells Anatole France's story about Barnabas, a poor street juggler who, after becoming a monk, performed his juggling act before an altar's image of Mary. "If I could juggle," Buckley adds, "I'd do so for Our Lady. I suppose I am required to say that, in fact, I have here endeavored to do my act for her."

When Buckley declares that his church is "unique" in having a "vision that has not changed in two thousand years," he means it has never abandoned such funda-

mental doctrines as the incarnation, virgin birth, atonement, and resurrection of Jesus. But these are doctrines
shared by conservative and fundamentalist Protestants.
With respect to doctrines peculiar to Catholicism, the
Roman church has changed enormously.

This is especially true with respect to Mariolotry.
Mary's immaculate conception, for instance, is nowhere
mentioned in the Bible. The notion that Mary was born
without the taint of original sin was not infallibly declared
a dogma until proclaimed by Pope Pius IV in 1854. Thomas
Aquinas and other eminent medieval scholastics had vigorously opposed it. Aquinas even quoted passages from
the gospels in which Jesus treated his mother with
obvious disrespect. Let me cite them.

When Jesus began attracting enormous crowds with
his charismatic preaching, we are told that his mother and
siblings, unable to get near him because of the press, sent
a message saying they wanted to talk to him. Did he
respond by asking his followers to let them through? He
did not. Instead he stretched a hand toward the multitude
and said, "Behold my mother and my brethren! For whoso-
ever shall do the will of my Father, which is in heaven, the
same is my brother and sister, and mother" (Matthew
12:46–50; Mark 3:31–35; and Luke 8:19–21).

When Jesus made his triumphal entry into Jerusalem
astride a donkey, a woman approached and cried out,
"Blessed is the womb that bare thee, and the paps which
thou has sucked" (Luke 11:27). Did Jesus thank the
woman for this tribute to his mother? He did not. Instead
he replied, "Yea, rather blessed are they that hear the word
of God and keep it" (Luke 11:28).

During the marriage ceremony at Cana, we are told (John
2:1–4) that Mary approached her son to tell him they had run
out of wine. Jesus rebuked her: "Woman [not mother], what
have I to do with thee? Mine hour is not yet come."

After his resurrection Jesus appeared to his disciples,
even to a prostitute named Mary, but not to his mother.

There is not a passage in the New Testament in which Jesus expresses love for his mother or even mentions his father. On the cross when he asks John, the disciple he especially loved, to take care of his mother, he did not say, "Mother, behold thy son!" He said, "Woman, behold thy son!" (John 19:26).

Saint Paul, whose letters predate the gospels, never mentions the virgin birth. His only reference to Mary is when he speaks (Galatians 4:4) of Jesus having been "born of a woman." Almost all Biblical scholars now agree that Jesus knew nothing about a virgin birth, and that this myth was a late addition to the gospels.

The Roman church has never ceased elevating Mary ever higher toward the status of a goddess. In 1950 Pope Pius XII infallibly proclaimed the doctrine, also nowhere in the Bible or in the works of early church fathers, that Mary did not die a natural death. Like Enoch, Elijah, and Jesus, her physical body never moldered in a grave. It was carried directly to heaven. Today there is growing pressure by Catholics to persuade the Vatican to make Mary a coredeemer with Christ. I do not expect this to happen. If it does it will be a blunder greater than the condemnation of Galileo. It will forever kill all hopes of ecumenicism.

Buckley has always been fascinated by Catholic converts, especially those from Protestant backgrounds. To gather material for his book he brought together what he calls his "Forum"—thinkers who converted to Rome. To each he asked a series of sharp questions about their beliefs. He apologizes for not being able to include his friend Clare Boothe Luce ("one of the liveliest minds I ever knew"), but she was too ill. Convert Malcolm Muggeridge was also too sick, although Buckley later devotes a chapter of high praise to him.

The Forum consisted of Father Richard John Neuhaus and Father George Rutler, both former Lutheran pastors; Russell Kirk, the well-known conservative writer; Jeffrey Hart, a senior editor of Buckley's *National Review*; Ernest van den Haag, sociologist; and Wick Allison, former

National Review publisher. In their responses, all defend the great miracles of the Bible, including, above all, the bodily resurrection of the Lord. To my amazement Kirk even calls the Shroud of Turin "the most wondrous of all relics," and "possibly" the actual shroud of the crucified Jesus![2]

The strangest chapter in *Nearer, My God* is devoted to a lurid account of the crucifixion excerpted from an English translation of a five-volume life of Christ by Maria Valtorta (1897–1961), an Italian mystic. Titled *The Poem of the Man God*, Maria obtained her tens of thousands of details about the life of the Lord from visions which she firmly believed were channeled through her by the Holy Spirit. I had never before heard of this bizarre work. I hope never to hear of it again.

To Buckley's credit he does not accept Maria's visions as authentic, but he is so fascinated by how she handled the excruciating details of the Lord's suffering that he squanders eighteen pages quoting them.

> My decision, then, is that in the only book on the faith I will ever put together I don't want to deprive the reader of what I view, notwithstanding its crudity—perhaps because of it?—as in artful portrayal of the great historical event that preceded, and led to, the Resurrection, a depiction if not inspired by God, inspiring nonetheless.

It is difficult for me to imagine anyone finding Maria's account inspiring. I found it stomach turning. Here are two small segments:

> The sufferings are worse and worse. The body begins to suffer from the arching typical of tetanus, and the clamour of the crowd exasperates it. The death of fibres and nerves extends from the tortured limbs to the trunk, making breathing more and more difficult, diaphragmatic contraction weak and heart beating irregular. The face of Christ passes, in turns, from very deep-red blushes to the greenish paleness of a person bleeding to death. His lips move with greater difficulty because the overstrained nerves of the neck and of the head itself, that for dozens

of times have acted as a lever for the whole body, pushing on the cross bar, spread the cramp also to the jaws. His throat, swollen by the obstructed carotid arteries, must be painful and must spread its edema to the tongue, which looks swollen and slow in its movements. His back, even in the moments when the tetanising contractions do not bend it in a complete arch from the nape of His neck to His hips, leaning at extreme points against the stake of the cross, bends more and more forward because the limbs are continuously weighed down by the burden of the dead flesh.

Maria's vision becomes agonizingly maudlin:

And fainter and fainter, sounding like a child's wailing, comes the invocation: "Mother!" And the poor wretch whispers: "Yes, darling, I am here." And then His sight becomes misty and makes Him say: "Mother, where are you? Are you abandoning me as well?" And they are not even words, but just a murmur that can hardly be heard by her who with her heart rather than with her ears receives every sigh of her dying son. She says: "No, no, my son! I will not abandon you! Listen to me my dear . . . your mother is here, she is here . . . and she only regrets that she cannot come where you are. . . ."

In addition to his Forum converts, Buckley also displays great admiration for others who, as he informs us, the British like to say, "poped": John Henry Newman, Gilbert Chesterton, Heywood Broun, to name a few. Many pages are devoted to a debate between convert Father Ronald Knox, a leading Catholic apologist in England, and Sir Arnold Lunn, a prolific journalist who at the time was highly skeptical of Catholicism. Their exchange of letters, in which Lunn raised questions and Father Knox replied, was published in 1934 as a now forgotten book titled *Difficulties*. Later, Buckley and Lunn became good friends and skiing companions. Lunn's last book, before he died in 1974, is dedicated to Buckley and Pat.

Some of the excellent questions raised by Lunn were: How can free will be harmonized with God's complete knowledge of the future? How can irrational evil, such as deaths from earthquakes, be reconciled with an all-powerful, all-merciful deity? How can a church, supposedly guided by God, justify the horrors of the Inquisition? The sale of indulgences? How can the church justify failing to be in the vanguard that opposed slavery? Above all, how can the church defend the extreme cruelty of Jehovah? Here is how Lunn phrased one letter:

> If every part in the Bible is equally inspired, we are forced to identify the God of the Psalms, the God of Isaiah, and the God of St. John with the barbarous and anthropomorphic Jehovah—a god who is angry and who repents, a god who demands blood sacrifices, a god who approves of the murder of women and children, a god who, in brief, represents at every point the most complete contrast with the God of the New Testament.

Father Knox does his best to counter Lunn's strong objections. To the paragraph just quoted he replies that the Bible was written by men who were not divine, but only fallible transcribers. Passages in the Bible have no "plain meaning." This is why we need a divinely guided church to interpret verses which can be disputed. "The whole of the Bible is immune from errors," writes Knox, "but we must have a living church to interpret it correctly."

Buckley of course always sides with Father Knox. To everyone's surprise in England, two years after his debate Lunn converted to Catholicism and was received into the church by Father Knox! The story of his conversion is told in his 1933 book *Now I See*.

As a non-Christian theist, I thought Lunn raised powerful objections to the faith and that Knox gave feeble answers. This is especially true of their exchange on the notion that God will punish some sinners with everlasting torments in hell. Buckley expresses hope that hell may someday be empty, and he takes comfort from Knox's statement that no one will suffer in hell who doesn't deserve it. There is no indication that Buckley is troubled by the Pauline doctrine that it is not good works which qualify one for heaven, but a belief that Jesus is the Son of God and the world's savior. Paul even adds the proviso that to escape damnation one must also believe God raised Jesus from the dead! Nor is he troubled by the fact that for centuries his church taught that no one would enter heaven who had not been baptized a Catholic. There were even medieval instruments for baptizing babies in the mother's womb if it was suspected that the child might not survive birth.

Lewis Carroll, a devout Anglican, wrote a pamphlet bashing the doctrine of eternal punishment. He said that if he believed for a moment that Jesus had actually taught such a dreadful dogma he would instantly cease to be a Christian. Asked if that meant he thought even Satan would eventually be saved, he replied that he did indeed believe this.

On page 125 Buckley makes a somewhat similar vow:

> If, *per impossibile,* it were established that Christ did not rise, I would myself instantly enlist in the Judaic faith, whose heaviest burden would then be not Jonah and the whale, but the blemish brought on Judaism by the fake Jesus who went around citing the Old Testament as his patrimony, claiming he was God. I would then, in Judaism, still be united with the prophets, and settled down for the promised incarnation sometime in the future.

What strange remarks! If a Christian ever became convinced that the New Testament is fake history, Buckley thinks it reasonable that he or she should then retain a faith in the accuracy of Old Testament history! Is it possible Buckley actually believes that God once drowned every man, woman, and child, and all the land animals, except for one undistinguished family because he was angry with the humans he had created?

What does Buckley think, I wonder, about a God who

orders Abraham to murder his son? How would Abraham know it was God speaking and not Satan? If the story of Abraham and Isaac is an allegory, what message does it teach? What moral can he find in God commanding Moses to murder all the men, women, and children of a neighboring tribe, but keep the virgin girls as slaves? What does he make of a God who destroys Moses' nephews with lightning bolts, like an angry Zeus, because the boys failed to mix properly the incense for an animal sacrifice?

What about Jephthah's murder of his beloved daughter merely to carry out a stupid vow? If this was an evil deed, why does Paul speak (Hebrews 11:32) of Jephthah as a man of great faith? One would have expected Buckley to say that if he abandoned orthodox Catholicism he would be content with a more liberal version of Christianity, or perhaps a philosophical theism. To suppose that a Catholic, unable to believe that Jesus' tomb became empty, should turn to the vengeful, loutish Jehovah of the Old Testament is a thought that boggles my mind.[3]

Here are some Catholic beliefs that Buckley has never doubted: The essence of Christ's body and blood in the bread and wine of the Eucharist, papal infallibility, the incarnation, the atonement ("The best way to put it, is that God would give His life for us and, in Christ, did."), original sin, hell, the trinity, the virgin birth, the immaculate conception, the assumption of Mary, and the miracles of healing by Jesus, and today in such sacred spots as Lourdes.

On more peripheral matters, which seem to be the only ones ever covered by the media, Buckley objects to Vatican II's replacement of the Latin mass by the "heartbreakingly awful English translation." He even disapproves of the lifting of the ban on Friday meat eating. He follows his church in opposing abortion and birth control (except for the rhythm method), the ordination of women, and the refusal (at least for now) to allow priests to marry.

Although I finished reading *Nearer, My God* with the satisfaction of finally learning that Buckley is more conservative a Catholic than he is a political conservative, a number of nagging questions remain.

How literally does Buckley take the Genesis account of creation and the Fall? On Sunday, October 12, 1997, he was interviewed about his book on *Firing Line*. Asked where original sin came from, he replied that it came from the fall of humanity in the Garden of Eden, "when Adam ate the apple." He cited Chesterton's *Everlasting Man* as a reference. Does Buckley take the story of the Fall metaphorically? Genesis does not specify the kind of fruit involved, but does Buckley go along with C. S. Lewis in thinking that the curse of original sin arose because Adam and Eve ate some kind of forbidden fruit that grew on some kind of mysterious tree?

Does Buckley believe that Adam was formed out of the dust of the earth and Eve fabricated from one of Adam's ribs, or does he accept the evolution of human bodies? Nowhere in his book does he speak of evolution. Does he buy the prevailing view today among Catholic theologians—it is probably also the present Pope's view—that human bodies did indeed evolve from apelike ancestors, but there was a sharp discontinuity when God infused immortal souls into the first human pair or first pairs? If so, this entails belief that the earliest true humans were reared and suckled by soulless beasts. Pat Buchanan is on record as rejecting evolution altogether, as did Chesterton, Hilaire Belloc, and so many earlier Catholic apologists. It would be interesting to know where Buckley stands on this question.

Sir Arnold Lunn was also a vigorous opponent of evolution. In his book *Flight From Reason* (revised 1932), he bashes H. G. Wells, frequently quotes Chesterton, and has high praise for George McCready Price. Price was a Seventh-day Adventist self-styled "geologist" who believed the entire universe was created in six days, about ten thousand years ago, and that fossils are relics of life destroyed by Noah's flood. Lunn grants that trivial evolution may have taken place within species (such as cats and dogs);

however, "it is not only possible, but probable, that God created different species. . . ."

Lunn was a good friend and admirer of the pugnacious British journalist Alfred Watterson McCann whose book *God—or Gorilla* (1922) is perhaps the most foolish book on evolution ever written by a Catholic. It has great pictures of apes, and a photograph of what McCann claims is a fossil shoe print from the Triassic period!

As late as 1950 Pope Pius XII issued an encyclical letter which said that Catholics must believe that all humankind descended from Adam, not from a number of first parents. Original sin, he argued, "proceeds from sin actually committed by an individual Adam, and which through generations is passed on to all and is in everyone as his own." Apparently this is Buckley's opinion also.

We know Buckley believes in angels. Does he also believe in Satan and the other fallen anngels? The devil and his demons are curiously absent from Buckley's confessional. He must believe in demons because the gospels tell how Jesus cast them out of persons possessed by them. In one memorable incident 2,000 devils were forced by the Lord to abandon a naked man and enter into a herd of pigs. The poor pigs then ran into the sea of Galilee and were drowned (Matthew 8:25–32, Mark 5:1–16, and Luke 8:27–36.) Does Buckley think persons are demon possessed today, and that his church has the power of exorcism?

Although Buckley admits he is no theologian, one would have expected him at least to glance into books by Jacques Maritain and Etienne Gilson, or by such Protestant heavyweights as Karl Barth and Niebuhr. Has he investigated the theology of Hans Küng? He seems equally indifferent to the writings of such philosophical theists as Immanuel Kant, William James, Ralph Barton Perry, Miguel de Unamuno, and the contemporary John Hick.

In 1990 I heard Madonna interviewed on television. Asked if she was a Catholic, she replied, "Sort of." Although not a practicing Catholic, she added, "once a

Catholic, always a Catholic." Küng, Father Greeley, and millions of other Catholics around the world are "sort of" Catholics. There is nothing "sort of" about Buckley. Unlike politicians who pretend to be Catholics to get votes, Buckley is the authentic true believer. He means every word he says.

I put down *Nearer, My God* with unbounded admiration for Buckley's courage and honesty, and the depth of his piety. There is not a trace of hypocrisy in his book. I also came away with the sad realization that Buckley is guilty of what has been called the sin of willful ignorance. He has never considered it worthwhile to learn much about modern science or recent Biblical criticism, much of it by Catholic scholars. He has made little effort to think through the implications of his beliefs in the light of such readily available knowledge.

Nearer, My God is a tribute to the awesome power of a great religious tradition to undergo uncritical transmission from parents to offspring. It will be interesting to see how Catholic reviewers respond to this passionate book. I was surprised to note that the only blurb on the back of the jacket is from Charles Colson, a convert from the shenanigans of Watergate to evangelical Protestantism.

NOTES

1. Buckley's first book, *God and Man at Yale* (1951), was a scorching attack on Yale for having abandoned God. He was roundly criticized for not mentioning anywhere in his book that he was an ardent Catholic.

2. Buckley may share this view. An editorial in his *National Review* (September 27, 1981) concluded that the Shroud was genuine to "a degree of probability that would have impressed David Hume."

3. Last November, on Tim Russert's TV show to promote his book, Buckley reiterated his vow that if he ever became convinced that the Lord did not rise from the dead, he would convert to

orthodox Judaism because of his great "respect" for the Old Testament prophets and their expectation of a coming messiah! That Jesus "venerated" his mother, Buckley said, "is well known." Asked about eternal life, he reminded Russert that Jesus promised it to the two thieves who were executed with him. Of course he promised it only to the "good" thief. I can only conclude that Buckley has never bothered to read the entire Bible.

ADDENDUM

William Buckley responded to my review in the *Los Angeles Times Book Review* (May 3, 1998):

> You devoted considerable space to what was labeled a review of my book, *Nearer, My God*, but was really something more like a freestyle display of the reviewer's complaints against Christianity. I would not impose on you for space required to list Martin Gardner's problems, let alone seek to analyze them. But let me deal with a few of the points touched upon.
>
> Gardner says that I have "made little effort to think through the implications of [my] beliefs." What *can* he mean by this? The implications of my beliefs can't go further than that some people go to heaven and some people go to hell. And if I have not thought through these implications—or the beliefs from which they derive—then neither has any Christian, layman or scholar or cleric, who recites the Nicean Creed. If that is the case—that all Christians are insouciant about the meaning of their faith—the consequences are absolutely enormous. You should consider giving over an entire issue of the *Los Angeles Times* to dramatize Gardner's concerns.
>
> "They"—that's me and other Catholics—"believe that God continually reveals new truths to his church, and that popes are infallible when they announce new doctrines." What *is* Gardner talking about? Catholics (and most other Christian congregations) believe that Revelation ended with the death of the last apostle. So what "new" truth is he talking about? When did a pope announce a "new" doc-

trine? The three papal declarations under the rubric of infallibility were reaffirmations of very old doctrines. The church's role as exegete is continuing but has nothing to do with the enunciation of fresh doctrine.

Gardner goes on to say that in his opinion, assorted Christians do not believe in assorted Christian dogmas. He comes up with some names, but to repeat them would be to encourage scandal. Yet Norman Vincent Peale, whose name he gives, is dead, so we may speculate without intruding upon him: Did he believe in the risen Christ? I don't know, but if he did not, what has that got to do with my book, *Nearer, My God*? If in that book I were taking aberrant positions, Gardner might have pleaded their eccentricity by citing traditional views with which they collide. . . . He has great difficulties with these matters. He describes me at one point as ultra-orthodox. But my views are mostly those given by the Catechism of the Catholic Church, as promulgated only a few years ago by the Vatican. What's the big deal about a Catholic supporting Catholicism, whose tenets are as defined by the magisterium? And even if the reviewer could come up with 100 names of men and women who call themselves Catholics but don't believe in Catholic dogma, what is the point in the exercise, in a review of a book that is concededly orthodox? To prove that some people are hypocrites, ho hum? Would a book arguing the validity of the United States Constitution merit a half-acre of a reviewer's space to point exultantly to people who do *not* believe in the Constitution?

"In light of all this doctrinal fuzziness, I have wondered for decades about where Buckley stood. . . . When I saw on sale his thirty-ninth book, *Nearer, My God*, I bought it at once. Would I find out at long last, I hoped, exactly what sort of Catholic he was and is?" Forgive me, but I have to wonder about the real pain Gardner suffered from all these years of suspense. If he was thus tormented, he might have eased his pain by writing me. I could have satisfied his curiosity in a postcard.

Then comes a lot of Clarence Darrow versus William Jennings Bryan, how-could-Jonah-live-inside-the-whale stuff. My book treats the question of biblical confusion, paradox, interpretation. And I devote an entire chapter to the evolution of doctrine, leaning on the thought and analysis of John Henry Newman. I wonder if it ever occurred to the reviewer to ponder the meaning of faith? He approaches Christianity rather in the manner of TV's Detective Columbo: "Oh, yeah, ma'am, there is just one more question that's been kinda bothering me—you say Mary actually went up to heaven physically, her body intact, and wearing WHAT exactly? . . . Just asking." Did he expect me to disguise the manifest difficulties in religious faith? My book *dealt* with those difficulties. What proceeds from Christian faith is manifestly inaccessible to those who approach the subject as local government inspectors holding clipboards and looking for code violations.

And then, finally, the apodictic, wince-making manner. About my chapter telling Mary Valtorta's version of the resurrection he writes: "I had never before heard of this bizarre work. I hope never to hear of it again. It is difficult for me to imagine anyone finding Valtorta's account inspiring." That there should be other points of view Gardner finds inexplicable. The same week in which his review appeared I received a letter which made the following commentary on that chapter:

"Never, and I do mean never, have I read an account of the crucifixion that so riveted my attention. I immediately circulated it to my wife, son, daughter, and son-in-law, insisting that they read it with Good Friday in mind. Never have I read a confession of faith that rang more true, and, yes, humble than yours. It is clear you are not comfortable with what in the Protestant tradition we call 'testimonies.' Yet that is what your work is: a testimony whose humility is its strongest coinage." This letter came not from a fellow illiterate but from Harry Stout, Jonathan Edwards professor of American Christianity at Yale University.

This is relevant to record given that Gardner, whose review appeared six months after my book was published, wonders how the book is being received. It is in an eighth printing, perhaps validating the attention the editor was so kind as to give it.

William F. Buckley Jr.
New York

I replied:

Buckley's letter surely confirms my contention that, in contrast to liberal Catholics, he remains securely ultra-orthodox. It would have been helpful if he had answered some of my questions, such as telling us his opinions about evolution, the reality of Adam and Eve, the cruelty of Jehovah, and other nontrivial matters. I won't bother to repeat them again to Buckley on a postcard because he is free to answer them in another letter.

I am puzzled by Buckley's claim that his church never offers "fresh doctrines" but merely affirms old ones. As I said, such doctrines as the Immaculate Conception and Mary's Assumption are nowhere in what Buckley calls the "Revelation" that "ended with the death of the last apostle." They are not even in the writings of the early Fathers. They were later beliefs that gradually arose among laypersons until they were finally validated by a pope.

Obviously Buckley is not going to take seriously the views of such recent Catholics who have "thought out their faith" as Hans Küng and Father Greeley. From Buckley's perspective these liberals are heretics who in earlier ages would have been excommunicated, perhaps burned at the stake. Buckley wonders if I ever pondered the notion of faith. I have. He will find my pondering detailed ad nauseam in my own confessional, *The Whys of a Philosophical Scrivener.*

I'm not surprised that *Nearer, My God* has had eight printings. My book with no advertising or promotion had six printings, and of course I am nowhere close to Buckley in fame. Both our confessionals are small potatoes compared to the sales of any book by or about Billy Graham.

Let me add that in a way I admire true believers like Buckley and Chesterton more than I do Catholics like Küng who, after abandoning all the unique doctrines of their faith, lack the courage to walk out of their church.

For space reasons, a number of paragraphs were cut from my review. I have here restored them.

"Being Catholic always mattered more to him [Buckley]

than being conservative," is how Garry Wills puts it in his *Confessions of a Conservative* (1979). That book's first chapters describe Wills's meeting with Buckley when he (Wills), fresh from a Jesuit Seminary, joined the staff of Buckley's *National Review*. The chapters contain entertaining portraits of young Buckley and his wife, and the magazine's early editors.

I reviewed Wills's angry attack on the Catholic Church, *Papal Sins* (Doubleday, 2000), in the *Los Angeles Times Book Review*, August 6, 2000. In that review I wonder why Wills, having abandoned those doctrines unique to Catholicism, remains a practicing Catholic. Apparently because he wants to radically reform a church he loves, and because of nostalgia for a happy Catholic childhood. As expected, Buckley's *National Review* (July 17) blasted Wills in a review by Robert Royal titled "The Church of Wills."

I also mentioned in my review what is not well known. In 1999 Mortimer Adler, after a lifetime of defending neo-Thomism, finally converted and was admitted into the Catholic faith.